Geography and science in Britain, 1831–1939

Geography and science in Britain, 1831–1939

A study of the British Association for the Advancement of Science

CHARLES W. J. WITHERS

Manchester University Press

Copyright © Charles W. J. Withers 2010

The right of Charles W. J. Withers to be identified as the author of this work has been asserted by him in accordance with the Copyright, Designs and Patents Act 1988.

Published by Manchester University Press
Altrincham Street, Manchester M1 7JA, UK
www.manchesteruniversitypress.co.uk

British Library Cataloguing-in-Publication Data is available

ISBN 978 1 5261 1671 0 *paperback*

First published by Manchester University Press in hardback 2010

This edition first published 2017

The publisher has no responsibility for the persistence or accuracy of URLs for any external or third-party internet websites referred to in this book, and does not guarantee that any content on such websites is, or will remain, accurate or appropriate.

Printed by Lightning Source

For my family, without whom

Contents

List of illustrations	viii
List of tables	xiv
Preface and acknowledgements	xv

1 Introduction: histories and geographies of science   1
2 Geographies of civic science: the British Association at work   24
3 The science of geography in the British Association   66
4 The dominion of science and geographies of empire: the BAAS overseas, 1884–1929   103
5 Hierarchy, distribution, connection: geography as a science of the physical world   135
6 Measurement, exploration and ethnology: geography and the human sciences   165
7 Science, education and the 'crisis' in geography, 1910–*c.* 1939   198
8 Conclusion: the British Association, geographies of science and the science of geography   232

Bibliography   248
Index   272

Illustrations

Cover picture 'Science in the Sydney Gardens.' Scenes from the BAAS meeting in Bath, 1888. Source: *The Graphic*, 15 September 1888, p. 289. Reproduced by permission of the Trustees of the National Library of Scotland

1.1 The opening of the BAAS annual meeting in Birmingham, 1865. Source: *Illustrated London News*, 16 September 1865. Reproduced by permission of the Trustees of the National Library of Scotland　5

2.1 Location of BAAS annual meetings, 1831–1939. Source: BAAS Reports and MS. Dep. BAAS, Bodleian Library, University of Oxford　25

2.2 Location of BAAS Conference of Corresponding Societies, 1883–1929, and (inset A) the number of corresponding societies by date of foundation. Source: based on lists in R. M. MacLeod, J. R. Friday and C. Gregor, *The Corresponding Societies of the British Association for the Advancement of Science, 1883–1929* (London: Mansell, 1975)　27

2.3 'Ticket map' and admission card for the BAAS meeting, Edinburgh, 1892. Source: Bodleian Library, MS. Dep. BAAS 179. By permission of the Bodleian Library, University of Oxford　37

2.4 Depictions of excursion science from the 1865 Birmingham BAAS meeting. Upper: 'The Wrekin.' Lower: 'Through the woods to the Wrekin – "the pursuit of science under difficulties".' Source: *Illustrated London News*, 23 September 1865, p. 281. Reproduced by permission of the Trustees of the National Library of Scotland　45

List of Illustrations

2.5 Depictions of science as performance, display and sociability, from the BAAS meeting, Cambridge, 1845: lecture demonstrations by Professor Brontigny and, to the foot, a 'floralia' in the gardens of Downing College. Source: *The Pictorial Times* VI (No. 121), 5 July 1845, p. 12. Reproduced by permission of the Trustees of the National Library of Scotland   46

2.6 Problems of comprehension, gently satirised in *Punch*. 'Things one would rather have left unsaid. "Ach! Gracious laty, I hope zat my long Cherman Lecture on ze Boetical Aspects of ze Bliocene Beriod did not *bore* you very much zis afternoon?" "Oh, not at all, Professor Wohlgemuth. I don't *understand* German, you know."' Source: *Punch*, 2 October 1886, p. 126. Reproduced by permission of the Trustees of the National Library of Scotland   52

2.7 The geographer-explorer at rest. David Livingstone at rest over the Africa he did so much to discover. (In the accompanying text Livingstone is made to resent his own 'discovery' by H. M. Stanley.) Source: *Punch*, 24 August 1872, p. 77. Reproduced by permission of the Trustees of the National Library of Scotland   53

2.8 'Tableaux-Vivants from the History of Science.' Part of the theatrical display of moments of scientific importance at the BAAS meeting, Edinburgh, 1892. (That for geography portrayed Columbus's discovery of America in 1492.) Source: Bodleian Library, MS. Dep. BAAS 179. By permission of the Bodleian Library, University of Oxford   59

3.1 Sir Joseph Lister giving the presidential address at the BAAS meeting, Liverpool, 1896. Source: Bodleian Library, MS. Dep. BAAS 447. By permission of the Bodleian Library, University of Oxford   85

3.2 Portrait of Sir John Franklin. Source: *Scottish Geographical Magazine* 11 (1895), facing p. 329. Author's collection   90

3.3 'View of the city of Bath, from the south-east' in 1864. The city was the scene of huge crowds to hear David Livingstone and of heated debates over human science in the BAAS. As the editorial to this picture noted, 'Bath undoubtedly offered an attractive *locale* for an assembly of the devotees of science. In the physical features of the neighbourhood it presents more than one problem well worthy of, but still awaiting, solution. It seems to challenge scientific investigation.' Source: *Illustrated London News*, 24 September 1864, p. 309. Reproduced by permission of the Trustees of the National Library of Scotland   93

3.4 Henry Morton Stanley, 'discoverer' of David Livingstone (cf. Livingstone's own views on Stanley's discovery of him in 2.7 here). Source: *Punch*, 24 August 1872, p. 77. Reproduced by permission of the Trustees of the National Library of Scotland   94

3.5 Apportioning responsibility for attending BAAS science. 'Paterfamilias, being unable to dismember himself to attend the various lectures, divides his family into eight sections, A to H. He will receive their reports on their return.' Note the clear association between geography, section E – as the father issues the ticket for that section – and attendance by women. Source: *The Graphic*, 15 September 1888, p. 290. Reproduced by permission of the Trustees of the National Library of Scotland   95

4.1 Science for a youthful empire. The cover of the programme of the South African Association for the Advancement of Science in collaboration with the BAAS, 1905. Source: Bodleian Library, MS. Dep. BAAS 198, fol. 17. By permission of the Bodleian Library, University of Oxford   109

4.2 Science and national enterprise. The cover of the Ontario handbook issued in collaboration with the BAAS meeting, Toronto, 1924. Source: Bodleian Library, MS. Dep. BAAS 248. By permission of the Bodleian Library, University of Oxford   115

4.3 Excursion science at Niagara Falls, 1884. Source: *Illustrated London News*, 30 August 1884, p. 205. Reproduced by permission of the Trustees of the National Library of Scotland   122

4.4 BAAS delegates en route for the BAAS meeting, Toronto, 1924, on the SS *Caronia*. Source: Bodleian Library, MS. Dep. BAAS 447, fol. 145. By permission of the Bodleian Library, University of Oxford   123

List of Illustrations   xi

4.5 BAAS delegates encounter the 'Other': North American
     Indian natives on the BAAS trans-Canada excursion, 1924.
     Source: Bodleian Library, MS. Dep. BAAS 447, item 47. By
     permission of the Bodleian Library, University of Oxford   125

4.6 'New Zealand's Call to the Man of Science.' The cover of a
     promotional leaflet issued by the New Zealand authorities
     to the BAAS. The call never evoked a response from the
     BAAS: plans to visit New Zealand as part of the 1914
     Australia meeting were curtailed by the outbreak of World
     War I, and plans of 1934 were hindered by lack of finance.
     Source: Bodleian Library, MS. Dep. BAAS 230/18. By
     permission of the Bodleian Library, University of Oxford   128

5.1 Section D, Biology, at the BAAS meeting, Manchester,
     1887. Source: Bodleian Library, MS. Dep. BAAS 447. By
     permission of the Bodleian Library, University of Oxford   138

5.2 The hazards of Antarctic science, as illustrated on the title
     page of Joseph Hooker's botanical work undertaken whilst
     on the expedition led by James Clark Ross 'for the purpose
     of investigating the phænomenon of Terrestrial Magnetism
     in various remote countries, and for prosecuting Maritime
     Geographical Discovery in the high southern latitudes'.
     Source: J. Hooker, *The Botany of the Antarctic Voyage of
     HM Discovery Ships Erebus and Terror in the years
     1839–1843* . . . (2 vols, London: John Murray, 1844–
     1847), Vol. 1, title page; quote at p. v. Reproduced by
     permission of the Trustees of the National Library of
     Scotland   144

5.3 The geographical distribution of plants, part of new forms
     of graphical depictions of natural relationships. Source:
     A. K. Johnston, *The Physical Atlas of Natural Phenomena*
     (Edinburgh: William Blackwood & Sons, 1850), plate xvi.
     Reproduced by permission of the Trustees of the National
     Library of Scotland   153

5.4 The global reach of British hydrographic surveying in 1876.
     Source: Lieutenant G. T. Temple, 'Hydrography past and
     present', *Report of the Forty-ninth Meeting of the British
     Association for the Advancement of Science* (London:
     John Murray, 1879), pp. 229–40, plate xv, facing p. 229.
     Reproduced by permission of the University of Edinburgh
     Centre for Research Collections   154

6.1 Ethnographic map of Europe in 1850, by the German ethnographer Gustav Kombst. Source: A. K. Johnston, *The National Atlas of Historical, Commercial and Political Geography* (Edinburgh: John Johnstone and W. & A. K. Johnston, 1843), plate 46. Reproduced by permission of the Trustees of the National Library of Scotland   167

6.2 Cartoon of Sir Roderick Murchison chairing the BAAS meeting, Birmingham, 1865. In part, the accompanying text reads: '*Geography*. The question "Where are you going on Sunday?" satisfactorily answered. This included practical demonstrations in Street Gymnastics, or the Use of the Globes and Poles.' Source: *Punch*, 23 September 1865, p. 113. Reproduced by permission of the Trustees of the National Library of Scotland   172

6.3 BAAS ethnology in the field. The topmost image (No. 5) shows Michael Connelly, Inishmann, described by the ethnologists as 'A burly man, with the largest head measured in the Middle island'; No. 6 (centre) shows 'A characteristic group of the young men of Aranmore'; and No. 7 (at foot) Michael Flaherty and two women, Inishmaan. 'Flaherty refused to be measured, and the women would not even tell us their names.' Source: A. C. Haddon and C. R. Browne, 'The ethnography of the Aran Islands, County Galway', *Proceedings of the Royal Irish Academy* II (1891–1893), pp. 768–830, plate XXIII. Reproduced by permission of the University of Edinburgh Centre for Research Collections   179

6.4 Celtic physiognomy and 'racial types'. These images were among many taken during BAAS-associated ethnographic fieldwork in west Ireland in the 1890s. Source: C. R. Browne, 'The ethnography of Garumna and Lettermullen in the County of Galway', *Proceedings of the Royal Irish Academy* V (1898–1900), pp. 223–68, plate IV. Reproduced by permission of the University of Edinburgh Centre for Research Collections   180

6.5 Portrait of Marion Newbigin, BAAS section E president, 1922. Source: *Scottish Geographical Magazine* 50 (1934), facing p. 257. Author's collection   188

List of Illustrations xiii

6.6 'My first gorilla.' Paul du Chaillu made much of his African explorations in his BAAS and other public talks. The gorilla looks startled, or is about to break into song. Source: P. du Chaillu, *Explorations and Adventures in Equatorial Africa* (London: John Murray, 1861), facing p. 71. Reproduced by permission of the Trustees of the National Library of Scotland 190

6.7 Louis de Rougemont, geographical hoaxer. De Rougemont's talk to section E at the 1898 Bristol BAAS meeting occasioned great interest among the public present and scepticism among the professional geographers. Source: L. de Rougemont, *The Adventures of Louis de Rougement as told by Himself* (London: George Newnes, 1899), facing title page. Reproduced by permission of the Trustees of the National Library of Scotland 191

7.1 Photograph of Charles Close, president of section E in 1911. By courtesy of Colonel R. F. Arden-Close 205

7.2 Productions in regional science. The cover of the BAAS Handbook, Leeds, 1927. With this work Fawcett and colleagues in the Leeds geography department established a model for regional science and BAAS handbooks. Source: C. B. Fawcett (ed.), *General Handbook* (Leeds: Chorley & Pickersgill, 1927). Reproduced by permission of the University of Edinburgh Centre for Research Collections 220

7.3 Afternoon tea at the BAAS meeting, Leicester, 1933. Source: Bodleian Library MS. Dep. BAAS 448, fol. 18. By permission of the Bodleian Library, University of Oxford 224

7.4 Excursion science in the inter-war years. These delegates to the BAAS meeting in Norwich in 1935 are on an excursion to Cromer. Source: Bodleian Library, MS. Dep. BAAS 48, fol. 331. By permission of the Bodleian Library, University of Oxford 225

Tables

2.1	Attendance at BAAS annual meetings, 1831–1931	47
3.1	Grants made in support of the work of the British Association for the Advancement of Science, by section, 1831–1931	69
3.2	Grants made to section E (geography), 1852–1936	72
3.3	Content and number of geography papers given to section E meetings, by topic and decade, 1851–1933	78
3.4	Section E presidential addresses, 1858–1939: speaker and subject	81
4.1	Date of publication and title, BAAS handbooks for overseas meetings, 1884–1929	112
7.1	Statistics of BAAS section activities, 1885–1909	202
7.2	Categories and trends in geographical article publishing, 1912, 1927–1929	222

Preface and acknowledgements

This is a work about the geography of science and the science of geography and the connections between them. With regard to the first of these terms, I explore the urban locations and regional settings in which that leading institution for the promotion of science in Britain, the British Association for the Advancement of Science, was at work in its annual meetings in the century or so from its foundation in 1831 to the outbreak of World War II. In this period and through its annual meetings, the BAAS promoted science as a cultural good, a civic science for a civil and an increasingly imperial society, throughout the towns and cities of Britain and Ireland. In 1884 the BAAS undertook the first of its overseas meetings, gathering that year in Toronto. Six further overseas meetings were held. One of the sections by which the BAAS operated was section E, geography. Examination of the work of this section and the subject of geography as a civil science forms the basis of my second theme. The book deals with how geography in section E of the BAAS was understood and practised as a science, with geography's content in this context, and with the practitioners and audiences for geography as a civic science.

In bringing these two broad concerns together, the book is a study in the workings of one leading institution for the promotion of science in civic context with particular reference to one science, geography, and an attempt to illustrate how the geographical sites and social settings in which BAAS science worked affected its content and nature. Rooted as it is in work in the history of science, which has recognised the importance of science's geographical dimensions, and work in the history of geography, which has moved away from simple national narratives and unproblematic chronologies to address questions of social context and epistemological complexity, my broader aim is to contribute to conversations between these intellectual communities.

As with all such ventures, the book owes much to the stimulus afforded by many people, chiefly to those whose work is cited here, but also to those many people who, knowingly and unwittingly, have at various times contributed, criticised and commented upon the ideas discussed here. The project on which the book is based began life in 2005 as a funded proposal to the UK Economic and Social Research Council under the title 'Geography and the British Association for the Advancement of Science, 1831–1933' (ESRC ref. RES-000-23-0927). I am extremely grateful to the Council for its support in addressing what, to me then, seemed such an obvious gap in our knowledge about the geography of British science and the science of British geography in the 'long' nineteenth century that I began to doubt even my own claims as to the legitimacy of filling it. I owe my greatest debt to Diarmid Finnegan and to Becky Higgitt, who, at one time or another between 2005 and 2007, acted as the postdoctoral research assistants on this project. They were, quite simply, outstanding: all one could wish for in their energy, insight and thoughtfulness over the project and for the quality of mind they brought to the interrogation and interpretation of the material uncovered. I hope I have done justice to their endeavours in the hours spent in BAAS archives in Oxford and in London and in numerous other holdings.

The work involved study of the main BAAS archives, held in the Bodleian Library, Oxford, and in the BAAS offices in London, and I acknowledge the help received in Oxford particularly, and from Sir Roland Jackson, who was Chief Executive to the British Association for the Advancement of Science when the work began, what is now the British Science Association. The book does not necessarily represent the Association's views on its own history or geography. Additional financial support for study of the BAAS archival records overseas was secured from the British Academy, and I am pleased to acknowledge its kind support in this respect. Becky Higgitt overcame her fear of flying for the sake of science and spent time examining at first hand the material relating to the BAAS overseas. For permission to quote from material in their care I am grateful to staff in the Bodleian Library, the British Association for the Advancement of Science, the Royal Geographical Society (with the Institute of British Geographers), the Mitchell Library, Glasgow, Leeds University Library, the University of Cambridge Library, Bristol Record Office, the University of Birmingham Library, the National Library of Scotland, the School of Geography and Environment at the University of Oxford, the British Library, the National Library of Ireland, the Scott Polar Research Institute, Cambridge, the Yorkshire Philosophical Society, the Universities of Edinburgh and Glasgow and Imperial College London, the Adolph Basser Library of the Australian Academy of Sciences, Cape Town Provincial Archives, the National Archives of

Australia, National Archives Repository in Pretoria, Pietermaritzburg Archives Repository, the University of Toronto Archives, Western Cape Provincial Archives and Records Services, and Wits University Archive Services.

At one time or another since 2005 conference papers based on the materials examined here have been given at meetings in Canterbury, Boston, Chicago, Edinburgh, Manchester, Oxford and Toronto, and I have profited greatly from the criticisms and comments received. Other individuals have at times had to listen to me as I debated with them (and with myself) over the central arguments, or read versions of chapters: my particular thanks in this respect go to Innes Keighren, David Livingstone, Alvin Jackson, Geoff Swinney, Ron Johnston, Peter Meusburger, Bernie Lightman and the anonymous referees for the Press. Versions of Chapters 2 and 3 have appeared, respectively, in the *British Journal for the History of Science* ('Historical geographies of provincial science: themes in the setting and reception of the British Association for the Advancement of Science in Britain, *c*. 1831–*c*. 1939', 41 (2008), pp. 385–415 (with Rebekah Higgitt and Diarmid Finnegan)) and *Transactions of the Institute of British Geographers* ('Geography's other histories: geography and science in the British Association for the Advancement of Science, 1831–*c*. 1933', 31 (2006), pp. 433–51 (with Diarmid Finnegan and Rebekah Higgitt), and for this I acknowledge with thanks Simon Schaffer and Alison Blunt as respective editors, and Cambridge University Press and Wiley-Blackwell as publishers. I also acknowledge Becky Higgitt as lead author and the University of Chicago Press as publisher for their permission to draw as I have upon our jointly authored article in *Isis* on the BAAS and women audiences: 'Science and sensibility: women as audience at the British Association for the Advancement of Science', 99 (2008), pp. 1–27. For assistance with illustrations and permission to include as illustrations material in their care I am grateful to Ray Harris (for Figures 2.1 and 2.2), Chris Fleet and George Stanley and the National Library of Scotland, Tricia Buckingham and Richard Ovenden and the Bodleian Library, the University of Oxford and colleagues at the Centre for Research Collections at the University of Edinburgh Library. Colleagues at Manchester University Press have been all I could have wished for: supportive, encouraging, patient. My wife, Anne, has had to live with it for longer than she might have wished: to her, as well as to my family, this is with much love dedicated.

C.W.J.W.

1
Introduction: histories and geographies of science

Reviewing the fortieth annual meeting of the British Association for the Advancement of Science, held in Edinburgh in August 1871, a reporter for the city's leading newspaper the *Scotsman* recounted the impact of the week-long scientific gathering upon the city's civic spaces:

> The British Association for the Advancement of Science comes among us with an authoritative air and an appropriating touch which in ordinary circumstances or people would be looked upon as intolerably intrusive or amusingly audacious. We are compelled to give it an invitation, and when it appears at the preconcerted hour, it takes possession of us in the most overpowering style. It fills our streets with its finger-posts; it takes our Courts of Law to lounge in; it seizes our University, and fills it with hurry and high debate; it soliloquises in evening dress in our public halls; it prescribes the preachers in our pulpits; it makes itself easy in our drawing-rooms; it raises commotion in our kitchens; it descends to the depths of our cellars; and exercises itself in a great variety of liberties which we are not in the habit of permitting to anybody, and all as if it thought that we ought to consider ourselves particularly well off in being utilised by so potent and august a visitor.[1]

Even as he recognised that the advantages for 'men of science' were great – and the advantages to the public 'still greater' – the reporter nevertheless felt moved a few days later to remark that the effect more widely of the Association's practice of annual scientific visitations was distinctly uneven:

> No one can question the great scientific services rendered by the British Association. It has undoubtedly raised the country from a state of apathy into one of appreciation for scientific pursuits; but that form of action which has enabled it to produce a powerful temporary impression ought to be followed up by another better fitted for continuing and promoting the impression already produced. The Association may, in fact, be compared to [a] gigantic boa-constrictor, which takes one hungry meal a-year, and lies in a semi-dormant state during the rest of the period . . . . The energy

is magnificent, but, at the same time, discontinuous and spasmodic, and though the inhabitants of cities such as Dundee and Bradford may for once in a generation receive a visit from the Association, for twenty or twenty-five years they are left to grow up – and they do grow up – in ignorance of the very existence of this great peripatetic body.[2]

These comments speak to many of the issues concerned with understanding the nature and effect of the BAAS as it worked to advance science throughout Britain following its establishment in 1831. The fact that at its foundation the Association was determined to be peripatetic and provincial, meeting annually in different cities, means that it is possible to consider a geography of Association science, initially at the level of the city and, from there, at the scale of the nation. Why that city then? What did it mean to local citizens when visiting scientific delegates and their work took possession of a city? And what did it mean to visiting delegates, for what, exactly, did they do in the annual meetings? Was the experience different for Edinburgh in 1871 (which city had received the BAAS annual meeting twice before, in 1834 and in 1850) than for Dundee (which had welcomed the Association in 1867) or for Bradford (the agreed venue for the 1873 meeting)? The *Scotsman* reporter clearly thought so. Rather than simply chart the location of BAAS meetings over time, we might ask whether the geography mattered to the science. That is to say, for instance, how far – if at all – did the local circumstances of different urban settings influence the nature of the scientific affairs discussed there? What, quite, was the nature of the short-lived but nonetheless clearly consequential visitation of associational scientific meetings that, to take the 1871 Edinburgh case, helped fill the city streets with notices, the university's spaces with 'hurry and high debate', the town's drawing-rooms with chatter and recollection, even as, we may presume, second- or even third-hand word of the science, the scientists and of the audiences reached the domestic staff in kitchens and cellars. If, then, there is a nation-wide history and geography to be considered in recovering the experience of BAAS science done locally, so, too, there are questions about science's presence in different cities, in different sites within any one city and to do with the practices through which science was made and received in these annual meetings in their urban settings and civic spaces.

These and other such matters lie at the heart of my concerns to understand the workings of the British Association as a form of civic science in Britain and in Ireland between 1831 and 1939 and, of course, even farther afield, since the BAAS held several overseas meetings from 1884. But they speak to only one part of the problem examined here, namely, what was the geography of BAAS science? The other part may be as simply put: what was the science of geography in the BAAS? Since the BAAS was divided into sections, each section a science, what did being a scientific section mean in terms of how the BAAS worked to advance science? Geography was for most of its institutional life in the BAAS managed as section E. What did this mean for geography? What did contemporaries understand by geography? Did this subject understanding change over time? How was geography put to work within the Association's meetings? It is also legitimate to ask what contemporaries, and the BAAS, understood by the term 'science' since, relatively, this was a new cultural phenomenon at least in terms of its associational and sectional organisation – the word 'scientist' in its modern connotations of professional engagement with systematic institutionalised natural philosophy dates only from 1834.[3] Did geography share its concerns, subject matter and method with the other sciences? Indeed, since it is possible to consider that different geographical locations may have made for differing experiences between and even within BAAS meetings – potentially, place and practice conditioned and constituted science locally – we might likewise ask if geography as subject was altered in its nature, even in its content and meaning, by geography as setting.

This is not to propose a crude determinism in explanation of the making of science or of geography.[4] It is, rather, to examine the connections between geography and science as social practices in place, using the BAAS as the central example and the relations between the geography of science and the science of geography as the ordering principles. It is to highlight the fact that, as the *Scotsman* testimony hints, the audiences for science in different places may have interpreted different things of the subject – indeed, of any subject – as it was delivered and debated in one place or another. For just as associational science was a new phenomenon, and scientist a new term in the early 1830s, science was not one 'thing'. The term embraced subject identity and the epistemic processes by which such identity was secured (science as discipline and disciplining procedures). It embraced a variety of routes to knowledge made through certain procedures and practices (science as method). Science was also differently delivered and received through public lectures, laboratory or field work and printed pamphlets and to different audiences (science as dissemination). Science was also shaped by who it addressed and where – in university rooms, public halls, pulpits, private drawing rooms, or, even, in the newspapers (science as audience). Science also came to mark certain people who gained status for their command of these disciplines or for advances in science's practice in the eyes of appropriate audiences (science as credible practitioners, professional and amateur, even 'celebrity' figures). In these terms, and as the illustration of part of the opening events of the annual meeting in Birmingham in 1865 suggests, science was about social display, spoken authority and performance as well as the pursuit of new knowledge (Figure 1.1).

These issues have wider resonance than in study of the BAAS alone. Considerable attention has been paid in recent years by historians of science and geographers and others interested in the spatial nature of science's making and to the settings for scientific activity, the production and reception of science in place and science's mobility over space. Let me turn first to work on the geography of science, and then to studies on the science of geography in Victorian and early twentieth-century Britain, before returning to consider the BAAS and its mission and practices in more detail.

Historical geographies of science

Where once science was taken to be everywhere the same and to have presented certain knowledge divorced from the situated social conditions of its manufacture and reception – the contexts of its discovery and of its justification – it is now more generally recognised that science everywhere bears the imprint of local circumstances. Attention to the importance of locality has built upon longer-running recognition of science as social practice. These twin related facts – the first that science is not a standard enterprise and that it reflects local conditions in its making, cognitive content, mobility and reception, the second that science is a social construction, reflecting and directing particular social and political interests in those localities – together constitute the significant achievements of 'the spatial turn' in the history and geography of science in the last twenty years or so.

Just as Kuhn's *The Structure of Scientific Revolutions* (1962) was influential in the understanding of scientific knowledge as an ineluctably social phenomenon, so later works and authors have helped shape the nature and direction of the spatial turn in the historical study of science. Ophir and Shapin in 1991 provided an important initial survey of the place of science in local context. Smith and Agar's *Making Space for Science* (1998) was much influenced in its overall interrogation of the power of space and of place by the work of Foucault and his attention to the cultural contingencies of questions of site and territory. Golinski's *Making Natural Knowledge: Constructivism and the History of Science* (1997) criticised progressivist notions of science and considered, amongst other things, the place of science's production and the instruments and cultures of its representation.[5] Geographers working

Introduction

Figure 1.1 The opening of the BAAS annual meeting in Birmingham, 1865

on the spatial dimensions of science have likewise offered useful syntheses, most notably perhaps Livingstone's *Putting Science in its Place: Geographies of Scientific Knowledge* (2003) although others have contributed valuable overviews.[6]

The result of these endeavours from within the history of science and from within geography is that a variety of taxonomic schema are now at our disposal in thinking about the sites and social spaces of science's production and reception, the nature of its mobility over space and the cultures of its use between and within communities, practitioners and audiences. For Livingstone, moves toward 'the cultivation of a spatialised historiography of science', initially proposed under the headings of the regionalisation of scientific style, the political topography of scientific commitment and the social space of scientific sites, were later refined to advance site, circulation and region as the principal organisational categories for understanding science's venues, movement and variant areal expression.[7] Smith and Agar structured their authors' discussions of the situated shaping of scientific knowledge around themes of territory, of working classes, pastoral privileges, metropolitan spaces, and research sites.[8] Others have advocated a three-fold approach to the geography of knowledge – a static geography of place, a kinematic geography of movement and consideration of the dynamics of travel, or, simpler still, advanced a twofold conceptualisation of the places of science's making and reception (science *in situ*) and matters of science's movement over space (science in motion).[9] One survey of the connections between science and the city has addressed what the editors term 'an urban history of science'.[10] Rather than seeing the city as just a location for science, a 'setting' in perhaps its simplest sense, their attention to the mutual constitution of science and cities addressed four themes: the rise of urban expertise; science and the representation of the city; places of knowledge and their urban context; and knowledge from the street.

The cumulative effect of these and other studies has been several-fold. One strong feature has been numerous site-specific narratives of science's making and reception. Studies have been made, for example, of botanical gardens, the laboratory, 'the field', museums, public houses, scientific institutions, country houses, even of ships.[11] Interests have focused on the instrumentalised replicability of science, on its mobility and standardisation.[12] Matters of scientific practice – of what, actually, scientists do – have also been addressed, highlighting the rhetorical, textual and embodied nature of claims to knowledge and issues of expertise (tacit and specialised), display, repetition and experimentation.[13] At the same time, attention has been paid to the different scales of science's production and to what 'local' means, since, in the view of two leading practitioners, the 'localist turn' in studies of science's making has tended to neglect bigger questions to do with the movement and success of science and the forms of its transmission. For the same reason, the differing reception of scientific knowledge has been evaluated and questions raised about the problems inherent in what Secord has called 'knowledge in transit'.[14]

We should not doubt, then, that this work has reinforced the importance of place to an understanding of the making and reception of science, and that it has helped establish news ways of thinking about science's mobility over space. At the same time, the result has also been to reject simplistic notions of science's 'production', 'reception' and 'mobility', to be cautious about localism as the necessary consequence of studies of science-in-geographical-context, and to insist upon the socially and spatially contingent nature of what science was held to be by looking at science's practices, across scales other than the local, at how it moved – in print, by word of mouth, in specialist outlets – not just that it did and for those reasons, to look at audiences as well as at practitioners.

This study of the geography of BAAS science from 1831 draws upon many of these conceptual concerns to do with site and local setting, different forms of scientific practice and dissemination, audiences and the 'making' of science at different scales from the local (in different cities) and the extra-local (in different civic spaces within cities) to the national and the trans-national. I want to know how and where the BAAS worked in its meetings, how it was received and what its several sciences set out to do. Before returning to the Association, however, it is helpful to review something of how the science of geography in Britain was understood in other ways and in different institutional spaces.

The science of geography in Britain, *c.* 1830–*c.* 1933

These dates mark, respectively, the foundation of the Royal Geographical Society (RGS), in London, and the establishment of the Institute of British Geographers, an initially small group of British academic geographers concerned at the relative lack of attention then being given to scientific research in that subject and what they saw as geography's still prevalent exploratory tradition. The foundation date of the RGS and its near coincidence with that for the BAAS is interesting given moves to scientific institutionalisation in that period more generally and, as we shall see, the connections between the RGS in particular and the BAAS were close. We may for convenience consider geography's place in British intellectual history and as part of the history of science in Britain in this period to have been apparent in two main ways: in geography's connections with empire and empiricism; and in geography's institutional and educational place. In neither sense has there been much consideration to geography's place as a form of civic science.

The foundation of the RGS helped formalise already existing connections between Britain's military institutions (principally the Admiralty, in later decades the army), and colonial and imperial imperatives to know one's own dominions. First-hand investigations of the world's shape and contents were undertaken with a view to its better mapping and improved knowledge of the peoples and resources in hitherto 'undiscovered' spaces. Geography as formulated in and through the work of the RGS, certainly in its first half century or so, took the form of exploration through expeditions – many of them to the polar regions as had been the case before 1830, or to Africa and to South America. The empirical practices of observation, survey and measurement went hand-in-hand with territorial advances in Britain's empire. Such issues were not alone the concern of the RGS: the Royal Scottish Geographical Society, founded in 1884, and several provincial geographical societies begun in the later nineteenth century, likewise had a strongly imperial agenda, the latter in being established in cities such as Hull and Manchester with local interests in wider imperial matters.[15] The intellectual connections between geography, empire and exploration long pre-date 1830 and the RGS.[16] But their heightened practical realisation in Britain from that date did much to associate geography with in-the-field empiricism, with exploration and with maps, progressive editions of which unproblematically represented additions to 'secure' knowledge as the geographer's hallmark and epistemic calling card.

Geography's 'modern' institutional and educational history in Britain, certainly from the later 1880s, was at one time considered largely synonymous with its development as a university subject. Accounts by contemporaries of geography's 'modernity' stress the role of geography in education and the foundation of teaching departments at Oxford (in 1887) and Cambridge (1888) as a definitive 'new' beginning for the subject.[17] These views have been echoed in some more recent studies.[18] But such claims to single foundational 'moments' have also been questioned by work pointing to the long-run presence of geography in Britain's universities, by studies of geography's status within British schools and of the strongly gendered masculinist historiography through which much British geography has been interpreted.[19] Many of the features characterising geography in nineteenth- and early twentieth-century Britain held true of other nations; the production of the 'discipline' of academic geography being in particular associated with the value of national territories and the worth of empires.[20]

In this and other work, scholars have exposed the disciplinary 'origins' of modern geography and revealed, for Britain especially, methodological connections between the history of geography and the history of science, not least, perhaps, given the impact of Livingstone's influential *The Geographical Tradition* (1992), which interpreted geography's history through the lens of the history of science by being attentive to local context, to science's social making and to competing narratives rather than to single narratives of lineal and progressive development.[21] What such work has also disclosed is how much more there is still to know about geography's history as a science: what contemporaries understood by its recognition as such, about the internationalisation of geographical discourse through international geographical congresses after 1871, for instance, or the marginal role of women in promoting the subject, the emergence of geographical methods, the history of geographical fieldwork, or geography's credentials as a laboratory science to name only a few topics.[22] Not least, of course, we know little about geography in the British Association for the Advancement of Science and its place as a form of civic and provincial science.

Placing the BAAS: towards an historical geography of associational civic science

Perhaps surprisingly given its foundation by many of the leading scientists of early nineteenth-century Britain, their shared concern to promote science as a cultural resource and to establish connections with their counterparts overseas, assessment of the BAAS has been uneven. Its early years in particular have been very well studied, the work of Morrell and Thackray proving especially valuable.[23] Other, edited, studies of this self-proclaimed 'Parliament of Science' have explored the Association's beginnings, its relationship to provincial publics through its annual meetings and for the later nineteenth century, its place in promoting scientific internationalism.[24]

In their 1981 work, Morrell and Thackray disclosed the close affiliations between BAAS members and the civic promotion of science and revealed the Association's democratic and meritocratic mission to have been compromised by the dominance, until at least the early 1850s, of aristocratic figures and other 'gentlemen' of science. Nevertheless, the BAAS gave organisational expression to the emergent identity of 'science' and of 'the man of science' in this period. Its principal functions were to facilitate communication among men of science and the promotion of science to the general public. Morrell and Thackray's attention to the working out of these functions and to the membership, ideologies, politics and utilities of science and to the attempts to establish an internationalist agenda in the BAAS before about 1845 is unlikely to be surpassed. Yet, as others have pointed out in discussing their work and that of the BAAS, the scientists who founded the British Association were at pains to ensure that their scientific pursuits supported natural religion, an attitude clearly supported by and supporting the predominance of liberal Anglican clergymen amongst its early leaders. But from the 1840s, this position was an awkward one in the face of new natural knowledge which challenged the precepts of theological explanation. The place of mankind in particular raised difficulties, and not just for the BAAS. Public engagement with, and political support for science changed during the course of the nineteenth century.[25] BAAS sections begun in the 1830s proved themselves not fit for purpose in later decades. Nor can we be certain that Morrell and Thackray's explanation of the pattern of BAAS meetings in academic and provincial locations before about 1845 necessarily holds for the BAAS at work in other places later in the century. The engagement of the working classes with science, and science's place in the market place of populist entertainment were both more marked in the later nineteenth and early twentieth centuries than before 1845. For these and other reasons, study of the BAAS after mid century is overdue, and its interpretation can now more readily draw on the conceptual insights of that work in the geography of science summarised above.[26]

Contemporaries' views of the BAAS varied. Those active within it saw the association as a means to promote science as a form of civil society, various presidents over the years extolling its virtues, significantly in support of established religion in the first twenty-five years or so.[27] Critics included Charles Dickens, who parodied the BAAS as 'The Mudfog Association for the Advancement of Everything'.[28] The supportive views of various officials in the Victorian period have their parallels in retrospective assessments by modern scholars: by Howarth, the BAAS's secretary upon its centenary, and by MacLeod and others writing upon its 150th anniversary.[29] For Howarth, one-time secretary to geography's section, section E, the BAAS's endeavours were unflaggingly monumental: 'To voluntary service in the interests of science the whole record of the British Association stands as one great memorial'.[30] MacLeod is more circumspect, seeing in the Association's history periods of success (from foundation to the 1870s) and periods in which it acted as a model for scientific bodies in the empire as an 'agent of British scientific diplomacy' (from the mid 1880s to c. 1905), but noting too periods of indecision (from 1905 to 1919), even of retrospection and uncertainty (1919 to about 1940), given public doubts in the inter-war period over the sought-for benefits of science.[31]

Worboys's study of several of the overseas meetings, mainly those in Canada, and Dubow's work on the BAAS South African meetings of 1905 and 1929 have highlighted something of how the BAAS worked in imperial context.[32] In pointing to the key role of science as a form of politics, and to BAAS international meetings as a means to promote local science as part of national agendas, their work affords an opportunity to examine the geographies of BAAS at transnational scales, to know how geography was put to work in these colonial settings, and to connect study of the association with research on the history of geography in terms of empire and empiricism.[33]

For several reasons, then, the BAAS lends itself to study in terms of the geographies of its science and in regard to geography as one of its sciences. Consider, for instance, the Association's annual meetings. Since the BAAS published an annual *Report* which summarised its work over a given year but did not publish a periodical, its programme of activities during the annual meetings was the effective point of contact between its constituent categories 'science' and the 'public'. Like others, I do not take these to be fixed and essential categories but, rather, things shaped, locally, in practice in different ways and to different purposes.[34] The Association worked in one sense through networks of correspondence between its members and the operations of its central committees. But where and how the Association really went to work in civic context was in and through its annual meetings.

The mapping of associational scientific activity is useful in identifying precisely where scientific pursuits took place and may profitably be pursued at a range of different scales. But simply plotting where in Euclidian space BAAS meetings took place is insufficient without recognising the variety of what took place in the located venues. Any distributional map of scientific practices invites scrutiny of how those 'truth spots' – as Thomas Gieryn characterises them[35] – came into being, how they shaped the knowledge produced in them, how they were occupied by human agents and what was the relationship that they helped sustain between physical and social space. BAAS meetings evolved a more or less standard format: a presidential address, sectional programmes of papers (themselves commonly begun by that section's presidential lecture), public lectures, 'Working Men's Lectures' or, in several of the overseas meetings, 'Citizen's Lectures', a *conversazione* or other formal social occasion such as a ball, and day trips or longer excursions. The fact that public lectures and 'Working Men's Lectures' were so titled points to different audiences within the BAAS's civic mission. Excursions typically went to local sites associated with particular sciences – factories for the chemists of section B, geological outcrops for the geologists of section C, for example – but also included more general sites of display such as exhibitions of scientific equipment or sites open to the public, in appropriate civic venues, but usually allied with particular scientific subjects (such as botanical or zoological gardens).

In these terms, consider moments within the BAAS meetings in Oxford in 1860 and in Belfast in 1874. In Oxford in 1860, the celebrated confrontation between Thomas Huxley and William Wilberforce, the first an advocate of Darwinian thinking, the second its staunch opponent, made the Oxford meeting memorable. The exchange occurred following a paper by John William Draper during a meeting of section D and was, thus, neither part of the presidential address nor of the sections' officers. Yet reaction at the time and afterwards was heated and enduring, much more complex than is presented by portrayals of the event as an iconic clash between science and religion.[36] In Belfast in 1874, physicist John Tyndall's presidential address established his reputation as a materialist and prompted great controversy over science's materialist base. If the Oxford meeting was a heated exchange which tested then established social conventions in debates over science and religion, Tyndall's Belfast address was tantamount to a declaration of war on theology: the controversy did not subside for years.[37] As James notes, 'By the early twentieth century, along with the 1874 British Association Belfast presidential address by John Tyndall (1820–93), the events of 1860 had come to be seen as one of the milestones in the process of the transformation of natural philosophy into natural sciences free from theological fetters.'[38]

Key moments in time in the longer complex relationship between science and religion, in the BAAS and outside it, were also key events in place: within certain towns, and as presidential, sectional or other addresses in front of public audiences. Precisely because of the differing reactions to the addresses, these events point not to the dissemination of scientific certainty from a single 'truth spot' so much as to the variable contours of science's making and afterlife, the reactions within one city not being the same as another, or shared by all who heard it or read about it afterwards. Moreover, these were spoken addresses in particular social spaces. As Lenoir has argued, considering the institutionalisation of scientific work outside the laboratory provides us not just with 'a vantage point from which to produce a contextualised, historically sensitive account of the production of scientific knowledge', it also shifts our analytic gaze to questions of practice and to the more mundane and contingent nature of science's making and reception.[39]

The BAAS moved its annual meetings each year, and in that simple sense its science was mobile, its politics predicated upon institutional mobility. But in its presidential addresses and in the many session papers, its science was mobilised through speech and in subsequent discussion. Speech is critically significant in explaining how scientific claims travel from person to person and from place to place, which is why the role of conversation in the spread of scientific knowledge – both elite and popular – has come under increased scrutiny.[40]

BAAS addresses and papers thus lend themselves to interrogation not just for declarations of science's (and geography's) content, range and purpose, but as speech acts, moments of rhetoric in certain sites – church halls, university lecture theatres and so on – wherein science's orators may have been constrained by the setting in what they could say and how. Study of the BAAS may help illuminate the different ways of knowing identified by Pickstone and Knight and others: not alone, of course, since a whole range of literary and philosophical societies and other bodies provided speech sites and display venues for science.[41] Such work has pointed, amongst other things, to the importance of different sites within city space and to how much there is still to know regarding the relationships between civic science and civil society in historical urban context.[42] Yet in their blend of talks, excursions, social perambulation and, after 1874, use of handbooks, BAAS annual meetings combined elements of 'holding forth', 'science gossip', 'display' and 'ballyhoo'.[43]

We must, nevertheless, be circumspect. The sources available to document the reception of the BAAS are different in type and survive only variably well. BAAS official papers – committee minutes and so on – are uneven in their coverage: section E papers are particularly 'thin' for the later nineteenth century. Newspaper accounts provide a valuable record of local audiences' reaction, but afford no systematic means of discerning how meetings were regarded or how different sections' programmes were attended. To have a review of a meeting or section programme is not always to know what its audience made of it. Neither do newspapers – properly, journalists' reports – always make clear an individual's intentions. Editorial and other intentions could colour what was said of the Association and how and why it was said: commentators in the *British Critic*, for example, who spoke in 1839 of 'that great assembly of motley philosophers now menacing Birmingham with its presence' did so not just out of a belief that 'they [the BAAS] have tended to no good' but from a fear that the social mixing associated with the meeting was a danger to established religion.[44]

Even so, newspaper reports and other evidence are valuable in permitting distinctions to be made within such general terms as 'reception', 'audience' and, even, 'science'. Although we may treat such accounts as evidence of how BAAS science was seen, heard and engaged with by its audiences, being at a BAAS meeting and writing about it – whether in specialist journals, a newspaper, in one's diary or in letters to friends – was a form of scientific production *and* reception. In such epistolary practices and textual spaces, science was 'made' for the absent specialist and for members of the non-attending but reading public, and re-presented in summary and, perhaps, errant form. In much the same way, reading the BAAS handbook about one's host town or about scientific venues in the local area was simultaneously a form of scientific reception and a source of local civic pride. For yet others, having the book to hand to guide was part of the making of science through regional survey in different ways – walking to visit a described site, photographing things there, striking strata with hammers, collecting botanical specimens – which demonstrated local residential expertise in the education of the transient scientific visitor. At such a moment, who was the 'professional' scientist, who the gifted 'amateur'?[45]

Who were the BAAS's audiences? How did they engage with its sectional activities? Since daily or week-long tickets were relatively costly, the working class must be presumed to have had neither the finances nor the time. But attending the Association's meetings also depended upon a commitment to science itself or to one science more than another: capturing this commitment to the populism of civic science is notoriously difficult.[46] Given the concurrent nature of sections' programmes, BAAS visitors could and did move from one subject to another: consuming science required a sort of social and intellectual 'circuitry', as people moved from topic to topic, room to room, scientific display to *conversazione*. Indeed, attendance at BAAS meetings may not always have been about science but about participation in a civic social gathering, and to do with being seen by one's peers to be the sort of person who *ought* to take an interest in science, even if much of it, in certain sections in particular, was never engaged with or, if engaged with, understood. Here, there are connections with that work which has pointed to the social construction of space, to the ways in which social practices affirm spatial identity.[47]

Such claims are evident, for example, in the view of one reporter on the 1878 Dublin meeting. Speaking of what he called his 'perambulatory science' – his movement across the meeting as a whole – the reporter affords images of differential comprehension, social bustle and the limited oratorical capacities of BAAS speakers in their different speech sites:

Some orators were holding forth, under discouraging circumstances . . . only a few [of the audience] seemed to pay any attention, and many had on them a half-amused, half bewildered look of astonishment at finding themselves in such society at all. A geologist spoke about a mountain range, and revealed the marvels of its hidden mysteries in weird sentences, and young and old ladies and gentlemen lounged in and out again as ignorant as before.

In the evening *soirée*, by contrast, 'here science was popularised and beautified, everything brilliant and gay went side by side with the most marvellous inventions and the deepest research; and harmony, elegance, and grace added their charms to the brightness and animation of the scene.' His view was no doubt coloured by the fact that 'We were of those who got seats, and not amongst the struggling multitude who surged around the portals of Professor Huxley's room'.[48]

These differences in what was on offer and what was held to be science matter. Science in the nineteenth century especially was in the marketplace, a form of cultural good on display as never before.[49] 'Popular science' is a problematic analytic category of course.[50] And questions of 'populism' – that 'set of discourses and identities which extra-economic in character, and inclusive and universalising in their social remit' as Joyce has it[51] – do not always reveal themselves in evidence about who was at an Association meeting, and why. But whatever it was for actors rather more than for historians, popular science was often more popular than scientific: with illuminated lectures and managed displays, audiences could be dazzled by the showmanship more than informed by the principles or, alternatively, and for women especially, be made only to engage with static displays of appropriate sciences such as botany.[52] Because of such assumed differences – and the fact, that, on occasion, they went in numbers to BAAS lectures to hear such figures as John Herschel, Michael Faraday, David Livingstone and Thomas Huxley – women were often represented as being either especially interested in or bored by certain subjects, with the effect that organisers constructed the programme to cater for such perceptions. But sectional differences complicate this picture. Botany, geology and as we shall see geography were considered appropriate female topics. Section A was reckoned hard and masculine. Section F and, from 1901, section L (education), attracted a large number of women in their audiences and, proportionate to other sections, as speakers, but the inclusion of women there and the topics of social and welfare reform that they articulated led many to see such topics as not suitable to the BAAS.[53] As we shall see, section E was, for good reason, often identified as the 'Ladies' Section'.

What follows examines these and other issues relating to the science in the BAAS: as a section, a subject, a method, a popular thing, a profession and as a set of cultural practices in place. Elaborated upon, these are the issues making up the geography of BAAS science. What follows looks also at the science of geography in the BAAS, one science among several: at its practical articulation in presidential addresses and paper sessions, at its content, methods and relationships with other sciences, at the professionalisation and professionalism of its practitioners, and at its audiences. In both respects, my approach to science and to geography is to see them as the construction and practice of a certain form of culture, civic and intellectual, thus to expose the contingent and located character of their production and reception.[54] In combination, my aim is to suggest that the historical geography of BAAS science can be used to explore what Warwick in *Repositioning Victorian Sciences* (2006) terms that 'knot of relations between place, practice and audience' and to see how the geography of associational civic science in the period 1831–1939 worked within British civil society at local, national and trans-national scales.[55]

Notes

1 [Anon.], 'The British Association in Edinburgh', *Scotsman*, 3 August 1871.
2 *Scotsman*, 9 August 1871.
3 The term 'scientist' as a general term by which the students of natural philosophy could describe themselves was first coined in 1834 by William Whewell as part of his review of Mary Somerville's book *On the Connexion of the Physical Sciences*: see R. Yeo, *Defining Science: William Whewell, Natural Knowledge and Public Debate in Early Victorian Britain* (Cambridge: Cambridge University Press, 1993), pp. 110–11; R. Yeo, 'Scientific method and the rhetoric of science in Britain, 1830–1917', in J. Schuster and R. Yeo (eds), *The Politics and Rhetoric of Scientific Method: Historical Studies* (Dordrecht and Boston MA: Kluwer, 1986), pp. 259–97. In terms of the focus of this book, it is worth noting Schaffer's observation that much of Whewell's early conceptualisation of science was informed by geographical – perhaps more properly, topographical – metaphors: S. Schaffer, 'The history and geography of the intellectual world: Whewell's politics of language', in M. Fisch and S. Schaffer (eds), *William Whewell: A Composite Portrait* (Oxford: Clarendon Press, 1991), pp. 201–31.
4 This is the argument of Harold Dorn: H. Dorn, *The Geography of Science* (Baltimore MD: Johns Hopkins University Press, 1991). For commentary on this determinist view, see D. N. Livingstone, *Putting Science in its Place: Geographies of Scientific Knowledge* (Chicago: University of Chicago Press, 2003), pp. 1–16.

5 A. Ophir and S. Shapin, 'The place of knowledge: a methodological survey', *Science in Context* 4 (1991), pp. 3–21; C. Smith and J. Agar (eds), *Making Space for Science: Territorial Themes in the Shaping of Knowledge* (Basingstoke and New York: Macmillan, 1998); J. Golinski, *Making Natural Knowledge: Constructivism and the History of Science* (Cambridge: Cambridge University Press, 1998).

6 Livingstone, *Putting Science in its Place*. I think also of S. Naylor, 'Introduction. Historical geographies of science', *British Journal for the History of Science* 38 (2005), pp. 1–12 (which introduces a theme issue of five papers on this topic); C. W. J. Withers, *Geography, Science and National Identity: Scotland since 1520* (Cambridge: Cambridge University Press, 2001), pp. 1–28; D. Finnegan, 'The spatial turn: geographical approaches in the history of science', *Journal of the History of Biology* 41 (2008), pp. 369–88; R. Powell, 'Geographies of science: histories, localities, practices, futures', *Progress in Human Geography* 31 (2007), pp. 309–30; P. Meusburger, 'The nexus of knowledge and space', in P. Meusburger, M. Welker and E. Wunder (eds), *Clashes of Knowledge: Orthodoxies and Heterodoxies in Science and Religion* (Heidelberg: Springer, 2008), pp. 35–90; P. Meusburger, J. Funke and E. Wunder, 'Introduction. The spatiality of creativity', in P. Meusburger, J. Funke and E. Wunder (eds), *Milieus of Creativity: An Interdisciplinary Approach to Spatiality of Creativity* (Heidelberg: Springer, 2009), pp. 1–10; P. Meusburger, 'Mileus of creativity: the role of places, environments and spatial contexts', in Meusburger, Funke and Wunder (eds), *Milieus of Creativity*, pp. 97–154.

7 D. N. Livingstone, 'The spaces of knowledge: contributions towards a historical geography of science', *Environment and Planning D: Society and Space* 13 (1995), pp. 5–34; Livingstone, *Putting Science in its Place*.

8 This is the framework adopted by Crosbie Smith and Jon Agar in their *Making Space for Science*.

9 The three-part categorisation is taken from S. Harris, 'Long-distance corporations, big sciences, and the geography of knowledge', *Configurations* 6 (1998), pp. 269–305. The two-part taxonomies are taken from C. W. J. Withers, 'The geography of scientific knowledge', in N. A. Rupke (ed.), *Göttingen and the Development of the Natural Sciences* (Göttingen: Wallstein, 2002), pp. 9–18, and from Finnegan, 'The spatial turn'. For a further summary of work in the field, principally organised in relation to localisms and matters of laboratory practice, see Powell, 'Geographies of science'.

10 S. Dierig, J. Lachmund and A. J. Mendelsohn, 'Introduction: toward an urban history of science', in *Science and the City*, *Osiris* 18 (2003), pp. 1–19. This introduces a set of thirteen papers organised under the twin headings of 'Science and the Rise of Modern Cities', and 'Science and the City after 1900'.

11 For examples, and in the order in which they are given, see E. C. Spary, *Utopia's Gardens: French Natural History from Old Regime to Revolution*

(Chicago: University of Chicago Press, 2000); R. E. Kohler, *Landscapes and Labscapes: Exploring the Lab–Field Border in Biology* (Chicago: University of Chicago Press, 2002); S. Naylor, 'The field, the museum and the lecture hall: the spaces of natural history in Victorian Cornwall', *Transactions of the Institute of British Geographers* 27 (2002), pp. 494–513; A. Kraft and S. J. M. M. Alberti, '"Equal though different": laboratories, museums and the institutional development of biology in late Victorian northern England', *Studies in History and Philosophy of Biological and Biomedical Sciences* 34 (2003), pp. 203–36; A. Secord, 'Scientists in the pub: artisan botanists in early nineteenth-century Lancashire', *History of Science* 32 (1994), pp. 269–315; S. Forgan and G. Gooday, 'Constructing South Kensington: the buildings and politics of T. H. Huxley's working environments', *British Journal for the History of Science* 29 (1996), pp. 435–68; D. Opitz, '"Behind folding shutters in Whittinghame House": Alice Blanche Balfour (1864–1936) and amateur natural history', *Archives of Natural History* 31 (2004), pp. 330–48; S. Schaffer, 'Physics laboratories and the Victorian country house', in Smith and Agar (eds), *Making Space for Science*, pp. 149–80; R. Sorrenson, 'The ship as a scientific instrument in the eighteenth century', *Science in the Field*, *Osiris* 11 (1996), pp. 221–36.
12 See the essays in M.-N. Bourguet, C. Licoppe and H. Sibum (eds), *Instruments, Travel and Science: Itineraries of Precision from the Seventeenth to the Twentieth Centuries* (London: Routledge, 2003) and K. Raj, *Relocating Modern Science: Circulation and the Construction of Knowledge in South Asia and Europe, 1650–1900* (Basingstoke and New York: Macmillan, 2007); A. Simoes, A. Carneiro and M. P. Diogo (eds), *Travels of Learning: A Geography of Science in Europe* (Dordrecht: Kluwer, 2003); S. Shapin, 'Placing the view from nowhere: historical and sociological problems in the location of science', *Transactions of the Institute of British Geographers* 23 (1998), pp. 5–12.
13 For example, B. Latour, *Science in Action: How to Follow Scientists and Engineers through Society* (Milton Keynes: Open University Press, 1987); K. Knorr-Cetina and M. Mulkay (eds), *Science Observed: Perspectives on the Social Study of Science* (Beverley Hills CA and London: University of California Press, 1983); A. Pickering (ed), *Science as Practice and Culture* (Chicago: University of Chicago Press, 1982); H. Collins and R. Evans, *Rethinking Expertise* (Chicago: University of Chicago Press, 2007). For the claim that the shift toward studies of expertise represents a 'third wave' in science studies (the first being the Kuhnian 'moment', the second associated with the social constructivist interpretation of science in society), see H. M. Collins and R. Evans, 'The third wave of science studies: studies of expertise and experience', *Social Studies of Science*, 32 (2002), pp. 235–96.
14 Ophir and Shapin, 'The place of knowledge'; Shapin, 'Placing the view from nowhere'; S. Harris, 'Long-distance corporations, big sciences, and the geography of knowledge'; J. Secord, 'Knowledge in transit', *Isis* 95 (2004), pp. 654–72. Secord has done much to establish the importance

of different geographies of interpretation and reception of science in his *Victorian Sensation: The Extraordinary Publication, Reception, and Secret Authorship of* Vestiges of the Natural History of Creation (Chicago: University of Chicago Press, 2000).
15 On British geography and the connections between empire and empiricism in this period, see M. Bell, R. A. Butlin and M. J. Heffernan (eds), *Geography and Imperialism, 1820–1940* (Manchester: Manchester University Press, 1995); the essays in A. Godlewska and N. Smith (eds), *Geography and Empire* (London: Blackwell, 1994); A. Crowhurst, 'Empire theatres and the empire: the popular imagination in the Age of Empire', *Environment and Planning D: Society and Space*, 15 (1997), pp. 155–74; F. Driver, *Geography Militant: Cultures of Exploration and Empire* (London: Blackwell, 2001); M. Jones, 'Measuring the world: exploration, empire and the reform of the Royal Geographical Society, c. 1874–1893', in M. Daunton (ed.), *The Organisation of Knowledge in Victorian Britain* (London: British Academy in association with Oxford University Press, 2005), pp. 313–36. On the role of the Royal Scottish Geographical Society, see Withers, *Geography, Science and National Identity*, pp. 195–235, especially pp. 198–210. On the provincial geographical societies, see J. M. MacKenzie, 'The provincial geographical societies in Britain, 1884–1914', in Bell, Butlin and Heffernan (eds), *Geography and Imperialism*, pp. 93–124.
16 L. Cormack, *Charting an Empire: Geography at the English Universities, 1580–1620* (Chicago: University of Chicago Press, 1997).
17 J. S. Keltie, *Report of the Proceedings of the Royal Geographical Society in Reference to the Improvement of Geographical Education* (London: John Murray, 1886); H. J. Mackinder, 'On the scope and methods of geography', *Proceedings of the Royal Geographical Society*, 9 (1887), pp. 141–60.
18 Endorsements of this view include P. I. Scargill, 'The RGS and the foundations of geography at Oxford', *Geographical Journal*, 142 (1976), pp. 438–61, and D. R. Stoddart, 'The RGS and the foundations of geography at Cambridge', *Geographical Journal*, 141 (1975), pp. 216–39.
19 On these topics, see C. W. J. Withers and R. J. Mayhew, 'Rethinking "disciplinary" history: geography in British universities, c. 1580–1887', *Transactions of the Institute of British Geographers*, 27 (2002), pp. 11–29; T. Ploszajska, *Geographical Education, Empire and Citizenship: Geographical Teaching and Learning in English Schools, 1870–1944*, Historical Geography Research Series Publication No. 35 (Cambridge: HGRG, 1999); D. Livingstone, 'British geography, 1500–1900: an imprecise review', in R. J. Johnston and M. Williams (eds), *A Century of British Geography*, pp. 11–44; R. J. Johnston, 'The institutionalization of geography as an academic discipline', in R. J. Johnston and M. Williams (eds), *A Century of British Geography*, pp. 45–90; R. Walford, *Geography in British Schools, 1850–2000* (London: Woburn Press, 2000); C. W. J. Withers, 'A partial biography: the formalization and institutionalization of geography in Great Britain since 1887', in G. S. Dunbar (ed.), *Geography: Discipline,*

Profession and Subject since 1870: An International Survey (Dordrecht: Kluwer, 2001), pp. 79–119. On the strongly gendered nature of the history of British geography, see C. McEwan, 'Gender, science and physical geography in nineteenth-century Britain', *Area* 30 (1998), pp. 215–24; A. Maddrell, *Complex Locations: Women's Geographical Work in the UK, 1850–1870* (London: Wiley-Blackwell, 2009).

20 M.-C. Robic, 'Geography', in T. M. Porter and D. Ross (eds), *The Cambridge History of Science*, Vol. 7, *The Modern Social Sciences* (Cambridge: Cambridge University Press, 2003), pp. 379–90; for discussions of geography, education and empire after 1870 in Germany, France, Italy, the Netherlands, Sweden, Russia, the United States and Canada, see the essays in Dunbar (ed.), *Geography*.

21 D. N. Livingstone, *The Geographical Tradition: Episodes in the History of a Contested Enterprise* (London: Blackwell, 1992).

22 I address some of these issues in C. W. J. Withers, 'Geography's intellectual traditions', in J. Agnew and D. Livingstone (eds), *A Handbook of Geography* (London: Sage, 2010 forthcoming). On fieldwork, see for example T. Ploszajska, 'Down to earth? Geography fieldwork in English schools, 1870–1914', *Environment and Planning D: Society and Space* 16 (1998), pp. 757–74; J. Camerini, 'Remains of the day: early Victorians in the field', in B. Lightman (ed.), *Victorian Science in Context* (Chicago: University of Chicago Press, 1997), pp. 354–77; F. Driver, 'Editorial. Fieldwork in geography', *Transactions of the Institute of British Geographers* 25 (2000), pp. 267–8.

23 J. B. Morrell and A. Thackray, *Gentlemen of Science: Early Years of the British Association for the Advancement of Science* (Oxford: Clarendon Press, 1981); J. B. Morrell and A. Thackray, *Gentlemen of Science: Early Correspondence of the British Association for the Advancement of Science* (London: Royal Historical Society, 1984).

24 R. MacLeod and P. Collins (eds), *The Parliament of Science: The British Association for the Advancement of Science, 1831–1981* (Northwood: Science Reviews, 1981); A. D. Orange, 'The British Association for the Advancement of Science: the provincial background', *Science Studies* 1 (1971), pp. 315–29; see also J. B. Morrell, *John Phillips and the Business of Victorian Science* (Aldershot: Ashgate, 2005).

25 R. MacLeod, *Public Science and Public Policy in Victorian England* (Variorum: Aldershot, 1996); F. M. Turner, 'Public science in Britain', *Isis* 71 (1980), pp. 589–608.

26 On these points, see T. W. Heyck, *The Transformation of Intellectual Life in Victorian England* (London: Croom Helm, 1982), pp. 61–5; J. Burchfield, 'The British Association and its historians', *Historical Studies in the Physical and Biological Sciences* 13 (1982), pp. 165–74; F. M. Turner, *Contesting Cultural Authority: Essays in Victorian Intellectual Life* (Cambridge: Cambridge University Press, 1993), pp. 48, 55, 183–5.

27 G. Basalla, W. Coleman and R. H. Kargon (eds), *Victorian Science: A*

Self-portrait, from the Presidential Addresses of the British Association for the Advancement of Science (Garden City NY: Anchor Books, 1970).
28 Cited in Heyck, *Transformation of Intellectual Life in Victorian England*, p. 62, and in C. A. Russell, *Science and Social Change, 1700–1900* (London: Macmillan, 1983), p. 187.
29 O. J. R. Howarth, *The British Association: A Retrospect* (London: British Association, 1931); R. MacLeod, 'Retrospect: the British Association and its historians', in MacLeod and Collins (eds), *The Parliament of Science*, pp. 1–16.
30 Howarth, *The British Association*, p. 264.
31 MacLeod, 'Retrospect', pp. 1–2.
32 M. Worboys, 'The British Association and empire: science and social imperialism', in R. MacLeod and P. Collins (eds), *The Parliament of Science*, pp. 170–87; S. Dubow, 'A commonwealth of science: the British Association in South Africa, 1905 and 1929', in S. Dubow (ed.), *Science and Society in Southern Africa* (Manchester: Manchester University Press, 2000), pp. 66–99.
33 See Driver, *Geography Militant*; Bell, Butlin and Heffernan (eds), *Geography and Imperialism*; Jones, 'Measuring the world'. The BAAS barely merits a mention in R. Stafford, 'Scientific exploration and empire', in A. Porter (ed.), *The Oxford History of the British Empire: The Nineteenth Century* (Oxford: Oxford University Press, 1999), pp. 294–319.
34 On the broad issue of the public understanding of science, I have found the following especially useful: R. Cooter and S. Pumphrey, 'Separate spheres and public spaces: reflections on the history of science popularization and science in popular culture', *History of Science* 32 (1994), pp. 237–67; J. Gregory and S. Miller, *Science in Public: Communication, Culture and Credibility* (New York: Perseus Publishing, 1998); S. Shapin, 'Science and the public', in R. Olby, G. Cantor, M. J. S. Hodge and J. R. R. Christie (eds), *The Companion to Modern Science* (London: Routledge, 1990), pp. 990–1001; S. Yearley, *Making Sense of Science* (London: Sage, 2005).
35 See T. F. Gieryn, 'Three truth spots', *Journal of the History of the Behavioural Sciences* 38 (2005), pp. 113–32; T. F. Gieryn, 'City as truth-spot: laboratories and field-sites in urban studies', *Social Studies of Science* 36 (2006), pp. 5–38.
36 A point stressed in F. A. L. James, 'An "open clash" between science and the Church? Wilberforce, Huxley and Hooker on Darwin at the British Association, Oxford, 1860', in D. M. Knight and M. D. Eddy (eds), *Science and Belief: from Natural Philosophy to Natural Science, 1700–1900* (Aldershot: Ashgate, 2005), pp. 171–93.
37 R. Barton, 'John Tyndall, pantheist: a rereading of the Belfast address', *Osiris* 3 (1987), pp. 111–34; Turner, *Contesting Cultural Authority*, pp. 270–1; D. Livingstone, 'Science, region and religion: the reception of Darwin in Princeton, Belfast and Edinburgh', in R. L. Numbers and J. Stenhouse (eds), *Disseminating Darwinism: The Role of Place, Race,*

Religion, and Gender (New York: Cambridge University Press, 1999), pp. 7–38.
38 James, 'An "open clash" between science and the Church?', p. 173.
39 T. Lenoir, *Instituting Science: The Cultural Production of Scientific Disciplines* (Stanford CA: Stanford University Press, 1997), p. 5.
40 A. Walters, 'Conversation pieces: science and polite society in eighteenth-century England,' *History of Science* 35 (1997), pp. 121–54; J. A. Secord, 'How scientific conversation became shop talk', in A. Fyfe and B. Lightman (eds), *Science in the Marketplace* (Chicago: University of Chicago Press, 2007), pp. 23–59; D. Livingstone, 'Text, talk and testimony: geographical reflections on scientific habits', *British Journal for the History of Science* 38 (2005), pp. 93–100; D. Livingstone, 'Science, site and speech: scientific knowledge and the spaces of rhetoric', *History of the Human Sciences* 20 (2007), pp. 71–98.
41 D. A. Finnegan, 'Natural history societies in late Victorian Scotland and the pursuit of local civic science', *British Journal for the History of Science* 38 (2005), pp. 53–72.
42 For example, the papers in L. K. Nyhart and T. H. Broman (eds), *Science and Civil Society*, *Osiris* 17 (2002); C. W. J. Withers and D. A. Finnegan, 'Natural history societies, fieldwork and local knowledge in nineteenth-century Scotland: towards a historical geography of civic science', *Cultural Geographies* 10 (2003), pp. 334–53; D. A. Finnegan, *Natural History Societies and Civic Culture in Victorian Scotland* (London: Pickering & Chatto, 2009).
43 J. V. Pickstone, *Ways of Knowing: A New History of Science, Technology and Medicine* (Manchester: Manchester University Press, 2000); D. Knight, *Public Understanding of Science: A History of Communicating Scientific Ideas* (London and New York: Routledge, 2006). My use of these four examples here is from Knight.
44 *British Critic*, 29 March 1839.
45 On this question for biology in Victorian Britain, see P. D. Lowe, 'Amateurs and professionals: the institutional emergence of British plant ecology', *Journal of the Society for the Bibliography of Natural History* 7 (1976), pp. 517–35; S. J. M. M. Alberti, 'Amateurs and professionals in one county: biology and natural history in late Victorian Yorkshire', *Journal of the History of Biology* 34 (2001), pp. 115–47; A. Desmond, 'Redefining the x axis: "professionals", "amateurs" and the making of Victorian biology – a progress report', *Journal of the History of Biology* 34 (2001), pp. 3–50; A. Secord, 'Science in the pub: artisan botanists in early nineteenth-century Lancashire', *History of Science* 32 (1994), pp. 269–315.
46 On questions of class and populism as the basis to social identity, see P. Joyce, *Visions of the People: Industrial England and the Question of Class, 1848-1914* (Cambridge: Cambridge University Press, 1991).
47 H. Lefebvre, *The Production of Space*, translated by D. Nicolson-Smith (Oxford: Blackwell, 1991).

48 *Irish Times*, 19 August 1878.
49 A. Fyfe and B. Lightman (eds), *Science in the Marketplace: Nineteenth-Century Sites and Experiences* (Chicago: University of Chicago Press, 2007); A. Fyfe and B. Lightman, 'Science in the marketplace: an introduction', in Fyfe and Lightman (eds), *Science in the Marketplace*, pp. 1–19; B. Lightman, *Victorian Popularizers of Science: Designing Nature for New Audiences* (Chicago: University of Chicago Press, 2007).
50 On this, see the essays under the sectional heading 'Historicizing "popular science"' in *Isis* 100 (2009). J. R. Topham, 'Introduction', *Isis* 100 (2009), pp. 310–18; A. W. Daum, 'Varieties of popular science and the transformations of public knowledge: some historical reflections', *Isis* 100 (2009), pp. 319–32; R. O'Connor, 'Reflections on popular science in Britain: genres, categories, and historians', *Isis* 100 (2009), pp. 333–46; K. Pandora, 'Popular science in national and transnational perspective: suggestions from the American context', *Isis* 100 (2009), pp. 346–58; B. Bensaude-Vincent, 'A historical perspective on science and its "others"', *Isis* 100 (2009), pp. 359–68.
51 Joyce, *Visions of the People*, p. 11.
52 G. Gooday, 'Illuminating the expert–consumer relationship in domestic electricity', in Fyfe and Lightman (eds), *Science in the Marketplace*, pp. 231–68; A. Shteir, 'Sensitive, bashful, and chaste? Articulating the *Mimosa* in science', in Fyfe and Lightman (eds), *Science in the Marketplace*, pp. 169–96.
53 For fuller discussion of these issues, see R. Higgitt and C. W. J. Withers, 'Science and sociability: women as audience at the British Association for the Advancement of Science', *Isis* 99 (2008), pp. 1–27.
54 These words consciously echo those of Lenoir in his 'approach to science as the construction and practice of a certain form of culture [which] highlights the radically historical, contingent, and local character of knowledge production': Lenoir, *Instituting Science*, p. 18.
55 The citation is from A. Warwick, 'Margins and centres', in D. Clifford, E. Wadge, A. Warwick and M. Willis (eds), *Repositioning Victorian Sciences: Shifting Centres in Nineteenth-Century Scientific Thinking* (London: Anthem Press, 2006), pp. 1–13, quote at p. 7. On civil society and associational activity, see the essays in S. Kaviraj and S. Khilnani (eds) *Civil Society: History and Possibilities* (Cambridge: Cambridge University Press, 2001), and G. Morton, B. de Vries and R. J. Morris (eds), *Civil Society, Associations and Urban Places: Class, Nation and Culture in Nineteenth-Century Europe* (Aldershot: Ashgate, 2006).

2

Geographies of civic science: the British Association at work

This chapter examines the British Association for the Advancement of Science at work in its annual meetings in different urban settings. Between its first meeting in York in 1831 – 'the most centrical city of the three Kingdoms', as one contemporary noted[1] – and the Dundee gathering of September 1939, which was brought to a premature end as war was declared against Germany, the BAAS met in thirty-seven different towns and cities in Britain and Ireland (Figure 2.1).

Several places were visited more than once. In keeping with its provincial mission, London was the site of the BAAS meeting only for the Association's centenary. Seven overseas meetings were also held: in Montreal (1884), Toronto (1897, 1924), Winnipeg (1909), South Africa (1905, 1929) and Australia (1914). My concern is to explore the connections between place, the location and the urban settings of Association activities, and the nature, conduct and reception of BAAS science. My focus is upon questions concerning the geographies of science – the places and the spaces in which it was conducted and engaged with – before turning to the science of geography.

The geographical location of BAAS meetings was not a matter of happenstance. The importance of different geographical locations for the annual meetings in helping realise the Association's objectives for the public promotion of science was a point stressed by the founding figures – 'It was then and there resolved that we were ever to be *Provincials*,' as John Dalton put it in 1831 – and it has been the subject of attention since. For Morrell and Thackray, having the annual meetings in different towns on a peripatetic basis meant that 'the manifest aim of advancing science could be fruitfully wedded to the latent function of social integration'.[2] Charting the geography of the early BAAS meetings (between 1831 and 1844), they distinguished a 'circuit of academic and metropolitan centres 1832–1835' (Oxford in 1832, Cambridge 1833, Edinburgh 1834, Dublin 1835), and a 'circuit of provincial towns 1836–1844' (Bristol 1836, Liverpool 1837, Newcastle 1838, Birmingham

Geographies of civic science 25

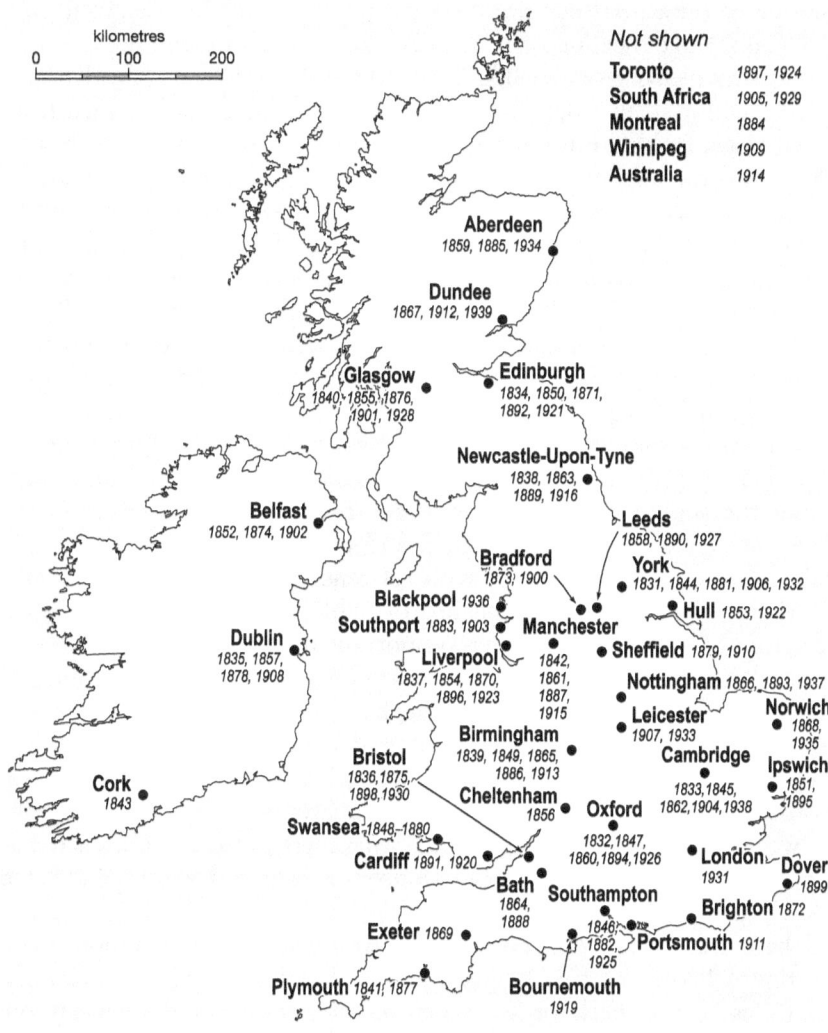

Figure 2.1 Location of BAAS annual meetings, 1831–1939

1839, Glasgow 1840, Plymouth 1841, Manchester 1842, Cork 1843 and York again in 1844). Neither the strategy of geographical movement nor the choice of locations was fortuitous: for these leading historians of the BAAS, 'The adoption of a provincial stance by the British Association did foster a vision of social integration, for it assuaged the pride of those peripheral groups represented in the rank and file and thus aided the gradual and complete take-over by the Gentlemen of

Science of the actual decision-making apparatus of the organization.'[3] The Cambridge meeting of 1833 was significant, for example, in establishing the identity of the BAAS and the influence of Cambridge academics within it. By 1835 the Association was dominated by men from Cambridge, Dublin, Edinburgh, Oxford and London. Only after the mid 1830s was the increase in provincial membership reflected in the visit of the BAAS to commercial and manufacturing towns, often prompted by resident scientific bodies in those places. The 1838 Newcastle meeting, for example, originated in proposals from local scientific bodies, as did Glasgow in 1840. For Morrell and Thackray, everywhere 'Local pride, civic rivalry, competitive emulation, and the desire for spectacle united to ensure participation, achieve harmony, and make manifest the resources of science.'[4]

If, then, as they further note, 'the British Association prospered as a philosophical travelling circus',[5] we should recall that these claims about the geographical mobility of the BAAS as a key feature in the Association's construction of itself, and of science for the public as 'a cultural resource', were made with reference to the years before 1845. Work on later periods has begun to highlight several features. Local scientific and other bodies were important in prompting an invitation to the BAAS, especially after the creation in 1883 of the Conference of Corresponding Societies, which brought together numerous local scientific societies, many of whom had already extended an invitation to the BAAS or been involved with it. In all, 162 local societies – of natural history, science, antiquarianism or combinations of them – were active between 1883 and 1929 (Figure 2.2). Members of these bodies acted as a network through which the BAAS promoted its endeavours and acted locally to assist and direct BAAS meetings.[6]

There was inter-urban competition to attract the Association, since hosting a BAAS meeting helped develop a town's social and scientific status as, for a week or so, scientists of national and international renown mixed with provincial figures, as *conversazione* and *soirées* provided social counterpoints to presidential addresses and formal academic sessions and as the public attended presentations, exhibitions and displays.[7] Similar features characterised the BAAS overseas, even as the Association additionally sought in that context to connect the imperatives of science and civic utility with those of politics and of empire, as we shall see in Chapter 4.[8] BAAS meetings did not simply move from one urban location to another. The competition between civic authorities to attract the Association, aided by leading men of science and local scientific organisations, was often stiff. Since scientists and others made use of the local area as a site for excursions or in other ways, where the

Geographies of civic science

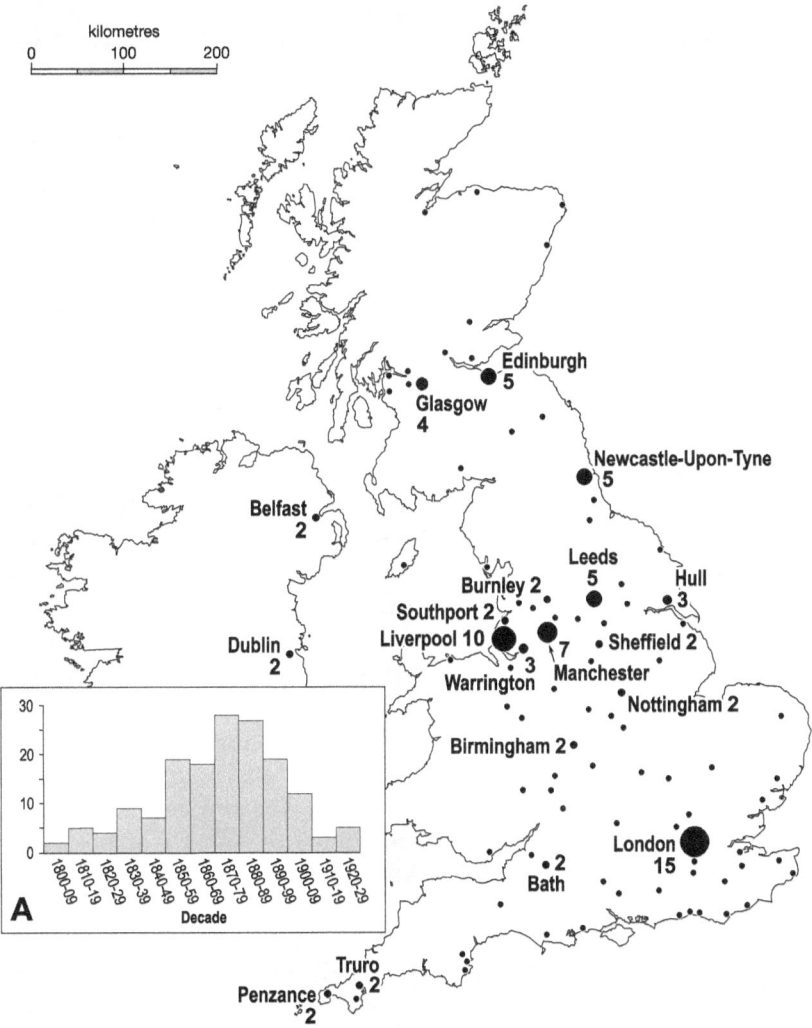

Figure 2.2 Location of BAAS Conference of Corresponding Societies, 1883–1929, and (inset A) the number of corresponding societies by date of foundation

Association met could and often did determine what sorts of things were discussed in its meetings, how and by whom. Both in general terms and in very specific ways – over the civic spaces available or not, for example, in terms of where sections held their papers, or in access to excursion sites – the nature of the place conditioned the sort of science undertaken there. It is possible, then, to do more than address the role of the BAAS in social and political terms, seeing its promotion of science as alone a 'cultural resource' and the emergence of a public agenda for science as the result of leading practitioners fighting to secure authority within the Association. Its meetings and its operating mechanisms can be examined in relation to their geographical setting to show how different local sites and social spaces were used by a body whose provincial agenda sought to advance science nationally and, after 1884, internationally.

With such a focus, different questions emerge about the nature of BAAS science, about its making and shaping in local context and about its reception across Britain and Ireland and farther afield. How did particular scientific societies or leading citizens influence the nature and focus of the science presented at BAAS meetings? How were given towns, their civic spaces and their surrounding localities used as venues for science? How was the BAAS received locally? Since there is evidence for some meetings that townsfolk accommodated visiting scientists at their own expense, we might expect there to be differences between the concerns of civic dignitaries to entice the BAAS, and locals' reaction to the actual arrival of visiting scientists and public delegates. There was certainly recognition at the Association's foundation that there would be different audiences – 'two interested classes', as was noted – 'visitors coming to meet their fellows in science and the intelligent persons of our own neighbourhoods who hope to be gratified and instructed'.[9] Did different local publics engage in the same ways with science as 'visiting fellows'? Questions about audience and about social and geographical differences are appropriate to ask, since most work on the Association's meetings and the production and reception of science in provincial context has tended to treat the BAAS as a whole. Almost no attention has been paid to the urban spaces utilised in its meetings or to the Association's constituent sections. And since BAAS meetings were both put together and managed differently by different practitioners, we must recognise the possibility that the science 'made' in any one place at any one meeting was engaged with differently by different audiences. Terms such as 'locals', 'the public' and 'audiences' have to be used carefully in considering how science was made, by and for whom and how it was received.

In addressing these issues, the chapter is in three parts. The first considers towns' and cities' civic promotion as appropriate venues for British Association science. The second explores the nature of BAAS annual meetings and considers how far the programmes – a mix of presidential addresses, section paper sessions, public lectures, excursions and other visits and social events – reflected local circumstances and where and how these elements of local civic science were undertaken.

The third pays attention to questions of practice and site by examining the reception of the Association and by considering the audiences' reactions to the conduct and content of Association science. My aim overall is to bring into sharper relief the means the Association used to realise its own geographical mission – and so to illustrate something of what is involved in studying the historical geographies of institutional civic science – before turning to look at geography as one section, one element of the BAAS mission for science and at practitioners' and audiences' views over what geography was.

Promoting and hosting provincial science: attracting the Association

Meeting places were chosen according to the views of BAAS officers. For them, the annual meeting should bring scientific and civic benefits to the place in question. They thus had to assess different civic invitations that stressed the scientific capacity of the location, the educational advantages for the local inhabitants, BAAS delegates and visitors and the financial support that local civic bodies could give the Association. Managing the invitations for a BAAS meeting required diplomacy: from the Association's officers, and from city figures in elaborating upon their prior written submission. During the 1838 Newcastle meeting, for example, the geologist Sir Roderick Murchison, then BAAS General Secretary, read out applications from civic authorities in Birmingham, Manchester, Glasgow, Sheffield and Hull, and an invitation to return to York. The president then 'invited any deputies present from the places which have sent invitations, to come forward to support their claims'. City officials from each of the towns did so. In the case of Glasgow, Baillie Paul was accompanied by two leading university professors (Nichol, Professor of Practical Astronomy; Thomson, Professor of Chemistry). For Hull, representatives exhibited a plan to show how and where the Association would be accommodated should it come. Discussion centred upon the respective merits of each, Birmingham being selected, as this was the city's second representation and because, later in the meeting, Glasgow 'stood aside' in its rival's favour. Having thus in 1838 secured what we might think of as an element of civic credibility, the eight-man deputation from Glasgow that again represented its case a year later in Birmingham in 1839 successfully drew upon its earlier graciousness in stressing then Glasgow's scientific credibility. Theirs was 'a city in which Science and the Arts have been so long, and so successfully cultivated, and on the advancement of which so materially depends the future commercial and manufacturing prosperity of the inhabitants'.[10] Glasgow was then the latest beneficiary of a policy which, as George Greenough (to warm applause) reminded delegates gathered in 1839, was about giving 'philosophers and men of science an opportunity of friendly intercourse with each other, and of distributing the knowledge which was to be found in London and Edinburgh throughout the whole country'.[11]

Such wording was commonplace. Civic dignitaries and leading officials regularly promoted their town or city as a scientific venue by reference to features there which should attract the Association. In 1847, W. R. Grove spoke on behalf 'of the Inhabitants of Swansea' in drawing to the attention of the Association's officers 'the peculiar local circumstances of the plan which might interest the Association, and the benefits which its visit might confer on the great mineral district of South Wales by a visit to Swansea'. His point was echoed in the support of the geologist William Daniel Conybeare, the Very Rev. the Dean of Llandaff, who wrote 'respecting the advantages to the advancement & diffusion of knowledge which might be justly expected from a visit to the Association to the great seat of our metallurgical operations'. Representations in support of Swansea had been made earlier. In April 1847, John Phillips had reported on the civic unity necessary to bring the BAAS there: 'Swansea is not so large a place, or so richly environed, as to be able to sustain a meeting of the Association except by the strength of united public feeling. This feeling is at present undoubtedly strong, and in the right direction, and there is no reason to believe that it will not remain so.'[12] On a motion proposed by Murchison and seconded by Professor Owen, Swansea's offer (initially made in Southampton in 1846) was accepted.[13] Quite why Conybeare, who lived near Cardiff, should support a meeting in Swansea is not clear: perhaps shared scientific interests in exposing the geological and mineralogical wealth of south Wales to others' scrutiny transcended local loyalties. Swansea was certainly the pre-eminent scientific town in South Wales in the 1830s and 1840s, given the presence of the Royal Institution of South Wales and the conjoint scientific interests of leading men of commerce and industry. (W. R. Grove was one of three FRS then active in Swansea.)[14] Likewise, the thirty-three signatories to the letter sent from Ipswich to the BAAS in August 1848 requesting 'a visitation' emphasised 'the interesting Geological Character of the locality, and the extensive Manufactures established in the Town' and hoped, moreover, that a visit 'would prove most conducive to the advancement of Science, and give an increased zest to that which already exists amongst the working classes, for the further investigation of Natural & Scientific phenomena'.[15]

While invitations stressed the scientific merits of the intended location (and, often, the scientific status of leading inhabitants), or gestured towards locals' intrinsic interest in science and drew attention to the mutual benefits that would accrue to the BAAS and to locals from a visit, the question of which to accept and, even, how to judge competing invitations was a source of concern to BAAS officers. In January 1848, the General Committee reiterated that it 'has the duty of appointing the place, time and officers of the Annual Meetings'. Further, 'By custom, this power has been limited to places which present invitations, to times suitable for those places, and to officers more or less indicated by local circumstances'. It additionally recorded that:

> The practice of obeying local invitations has been productive of good and evil: good by the spontaneous awakening of many important places to scientific activity; evil by the introduction of elements of display, temporary expedients, and unnecessary expense. These have somewhat impaired the efficiency of the Meetings, by withdrawing attention and consuming time which could ill be spared from the essential business of one scientific week.[16]

The fact of an invitation stressing local benefits did not always result in attendance by locals. 'By selecting for our place of meeting a central accessible point in an interesting district, where science has food and life, we may expect to secure a large local attendance of new members, and yet not lose our friends from a distance'. 'But it has happened' [recorded the General Committee] 'that a meeting by invitation has been so ill attended from public occurrences and local peculiarities, as to cause a loss ... to the Association Treasury'. In August 1848, John Phillips, then the Association's Assistant General Secretary, put before Council a discussion paper entitled 'Reasons for thinking that the Annual Meetings of the British Association ought not to be restricted to the places which present formal invitations and guarantees of expense'. Since the decision had already been made to hold the 1849 meeting in Birmingham, no further action was taken regarding Phillips' proposal.[17]

In order to prevent hurried planning and unnecessary competition between potential and competing locations, Council agreed that meetings were to be fixed two years in advance. Where there was doubt or dispute as to location, preference was given to places not before visited. In discussing the respective merits of Leeds and Manchester as sites for the 1858 meeting, for example, the towns were held to have more or less equal advantages. What finally swung the Association towards Leeds (that town being favoured only after an amendment proposed by William Whewell) was the fact that it was at the centre of 'a great district never yet visited by the British Association viz ... the largest district of the Kingdom uncultivated by the Association'.[18] Motivated by the fact that the BAAS had not met 'in the West-Riding of Yorkshire, a district which offered, in its natural resources and manufacturing industry, a wide field of interest', the Leeds Philosophical and Literary Society reflected in 1859 upon the 1858 Leeds meeting:

> The Local Committee believe that the benefits which it [the meeting] conferred on the town of Leeds were neither few nor trifling. It called forth a large amount of public spirit and of individual energy, and was the means of eliciting from several of our townsmen very valuable contributions to our stock of scientific knowledge. It awakened a new and lively interest in science and scientific men amongst considerable numbers of our population, and can hardly have failed to create in many cases a desire for more extensive and accurate information. It brought together the theorist and the practical man, who commonly move in separate and remote spheres. It established friendly and personal relationships with many of the distinguished leaders in science, from which the town has already reaped valuable results; and it made known to large numbers of the most educated class the true position of Leeds, as a seat of manufacturing industry and enterprize. The Meeting of the Association, however, they feel, should not be regarded primarily with reference to the benefit which it may have conferred on ourselves. Science is the foundation of the wealth and prosperity of Leeds and it was fitting that, when the opportunity offered, its citizens should welcome and honor the Masters of Science.[19]

What Whewell saw as filling a geographical gap consistent with BAAS strategy and some officers of the Leeds body interpreted as bringing scientific prestige to their town, others saw as a matter of appropriate civic environment. Discussions with a view to an invitation to, and civic deputation from Leeds had begun as early as 1850, invitations being renewed annually, without success, between 1851 and 1857. What held Leeds back was not shortage of local science nor shortfall in civic will but the lack of a proper building in which to host the BAAS. In the view of some within the Leeds Philosophical and Literary Society, the completion of the town hall – and not the fact that the region had not been visited by the BAAS – was the crucial factor in influencing the BAAS General Committee towards Leeds in 1857: 'It was pointed out that Leeds had long desired a visit, and now that the only difficulty in the way was clearly removed by the erection of the Hall, it was contended that our town was clearly entitled to take precedence of Manchester, which had already received the Association'.[20]

The BAAS was at pains to locate its meetings in settings where it thought benefits would accrue and to do so when civic spaces were available, not simply to return to established academic and industrial locations. Yet local bodies sometimes had to be persuaded. In his 1853 pamphlet *On the prospective advantages of a visit to the town of Hull by the British Association for the Advancement of Science*, initially given in November 1852 as an address to the opening session that year of the Hull Literary and Philosophical Society, Charles Frost presents what is perhaps the clearest case of a man of science trying to convince local others of the value of a BAAS meeting.[21] Frost (who was President of the Hull body) noted in general that Association meetings could stimulate science locally: 'Herein we have a proof afforded of the utility of the Association in calling into action native talent, and exciting such of the inhabitants as possess a taste for science, to qualify themselves in advance, for taking an active part in the preparation of the intellectual treat to be placed before their philosophical guests'. He noted specifically the benefits to tourism of the Association meeting, the improvement of local facilities (including new meeting rooms for the Literary and Philosophical Society), and the moral and intellectual benefits of association with visiting experts: 'It is scarcely possible to appreciate too highly the mind-purifying and soul-ennobling effects of coming into familiar contact with a vast assemblage of the master spirits of the age'.[22] Yet he was also circumspect. For Frost, BAAS meetings could act in isolation from their immediate civic setting unless care was taken to the contrary: 'In short, the Association, when in the height of its activity, may be compared to a little commonwealth, which has parasitically located itself in the midst of the visited town, and there acquired, for a brief space, a local habitation – with a population of its own – engaged only in its own pursuits – and governed only by its own peculiar laws and customs'.[23] He was not alone in this view.

By the 1860s and 1870s, the expressed conjoint interests of civic bodies, leading manufacturers and local scientific institutions and their representatives was an increasingly common feature of the deputations and written invitations made to the BAAS. Following 'a large and influential meeting of the principal bodies and Societies of this district', Newcastle's invitation to the Association in 1862, for example, stressed how facilities had changed since the 1838 meeting and highlighted the emergent industrial development of the region:

> The Scientific Interest of the neighbourhood of Newcastle has increased since 1838 in equal proportion, and new Branches of Trade have sprung up; amongst others the manufacture of the Metal Aluminium and Aluminium Bronze; and the smelting of Copper Ores; the manufacture of metallic Copper is also new to the Tyne. In Geology the great iron stone field of Cleveland has been discovered and practically worked.[24]

Having received earlier notice of Newcastle's intentions from Isaac Lowthian Bell, the city's mayor, John Phillips's reply hints at several of the issues being borne in mind in deciding meeting locations: 'There will be of course other claimants, but the ground is *all fair* for canny Newcastle; & as several later meetings have been far south of her, she has a clear locus standi, & the prestige of a never-to-be forgotten success. *One thing:* bring a delegation and documents and proof of *space in rooms* for a *large* meeting'.[25] Access to the right sort of civic space – in the form of a town hall, public lecture theatres and, as we shall see, museums for scientific and social display – was always on the BAAS agenda.

The development of sub-disciplinary identities and of science's institutional expression during the nineteenth century was reflected in the many provincial bodies which helped orchestrate bids to the BAAS and, if successful, then shaped Association science in given towns. In Norwich in 1868, for instance, the City of Liverpool represented its case for a future meeting through the mayor and two other leading figures from the Corporation, three representatives from each of the city's Literary and Philosophical Society, Historic Society, Polytechnic Society, Chatham Society, the Naturalists' Field Club and by delegates from the Chemists' Association, the Geological Society and the city's Medical Institution.[26] The fact that a meeting was held at all in Norwich owed much to the efforts of Sir Charles Lyell and others. During debates in the mid 1860s over the place of anthropology and ethnography and the nature of section E (discussed in Chapter 6), Lyell had noted how the BAAS still had constituencies in the country unvisited: 'There were parts of the country where there was no great wealth, such as Norwich, where there were a great number of students – isolated students of science – who, if they were brought together by the visits of the Association, might be numbered among them, and assist in the great work of scientific enquiry'.[27]

Liverpool's delegation to Norwich in 1868 had to compete with similar representation from Exeter, Edinburgh, Bradford and Brighton. The civic parties respectively made their case: that for Exeter centred on the Association not having been to the West of England; that for Edinburgh on preserving an order of precedence without justification in terms of any agreed Association rules: 'The deputation from Edinburgh expressed a readiness to give way to Exeter next year, as they had done to Norwich this year, provided Liverpool would give way to Edinburgh in the year following, and so preserve something like the order in which former meetings have been held, Edinburgh having on both occasions preceded Liverpool.' Edinburgh's case was rejected both because the Association aimed to 'distribute its meeting as impartially as possible over the United Kingdom; and that, as the meeting was held last year at

Dundee, the time had not yet arrived for another visit to Scotland'.[28] The Dundee meeting in 1867 here referred to had been strongly supported by numerous local borough councils.[29]

When, in Glasgow in 1876, competing invitations from Leeds and from Dublin were supported by numerous local scientific and literary institutions, a vote was taken, Dublin winning by a majority of twelve.[30] In turn, in Dublin in 1878, the Association's General Committee received and read out invitations from various bodies from Sheffield and beyond as that town presented its case: the Borough Council, the Corporation of Cutlers, the Town Trust, the Chamber of Commerce, Sheffield's Literary and Philosophical Society, its Naturalists' Club, from the Yorkshire Geological and Polytechnic Society, the Council of the Borough of Rotherham, Rotherham Literary and Scientific Society, the Yorkshire Naturalists' Union and from the Mayor, Aldermen and Burgesses of the Borough of Barnsley.[31] In 1896, Glasgow's delegation withdrew its intended invitation to the BAAS after finding, upon arrival in Liverpool, that arrangements for holding the 1898 meeting in Bristol were so advanced 'that it would have been neither courteous to that town nor proper to now present an invitation from Glasgow for that year'.[32] In contrast to the willingness with which academics at Cambridge had earlier welcomed the BAAS, views within the university in 1901 over a possible further visit were mixed: one member of the Philosophical Society there arguing 'Cambridge should restrict herself to international gatherings,' another complaining 'that I know it will murder a Long Vacn. as far as private work is concerned'.[33] Grudgingly, the Association was invited again to Cambridge, returning in 1904.

Such equivocation over the BAAS meetings had been apparent since the later 1880s. It was increasingly evident after 1914 as the Association was undermined by the growth of specialist meetings, by growing government funding for science and, in places, by the strength of local bodies some of whom saw little benefit in diverting scientific energies and financial resources towards a week-long programme of general science and away from their own initiatives.[34]

This evidence for the decisions and processes behind the location of BAAS meetings after 1845 suggests elements of continuity with the pre-1845 picture: the importance of local scientific bodies, of influential 'Gentlemen of Science', financial assistance (often promised, not always found), appropriate facilities (existing or planned) and the stimulus of locals' interests in science as a civic good, a stimulus often realised in the collective agency of local scientific societies with civic authorities. What is also true is that there was, from the second half of the nineteenth century, an increased recognition by urban authorities and local scientific societies that their own capacities, and, importantly, the significance of the local area, would attract the BAAS and could influence the content of meetings and of sectional programmes. At the same time, there was after 1850 or so a concern by the BAAS to meet where they had not before (as in Hull in 1853, Leeds 1858, Dundee 1867), not to visit the same area if another town had recently had a meeting (as happened to Edinburgh in 1868), and to return to towns when invitations, boosted by the success of a previous visit, could ensure appropriate new venues (Newcastle 1868) and/or the support of neighbouring civic authorities (Dundee 1867, Sheffield 1879, Liverpool 1896, Edinburgh in 1892 and 1921).

A clearer sense of how towns and cities were science's 'making' and 'selling places' is apparent if, in two senses, we change our scales of analysis: geographically, to look at particular urban sites and the uses made of the local area as scientific sights, and epistemologically, by considering the practices which the Association, its delegates and its audiences, drew upon to make science work locally.

Practising civic science: BAAS meeting programmes

BAAS meetings embraced scientific events such as presidential addresses, section sessions, lectures and displays, and social functions. Because this is so, it is not helpful to think of BAAS meetings as having any single and precise distinction between spaces of and for science, and spaces of social activity. Of course, lecture rooms were given over to talks and presidential addresses as sites for 'talking shop' and, usually, listening attentively.[35] But attending science lectures to hear speakers or viewing geological collections to marvel at nature's oddities was, for many, also an opportunity to view and to be viewed. Even the presidential address, ostensibly the most formal moment in the BAAS programme of the range and purpose of a science or declaration of science's utility, was an occasion for the audience to view one another, for experts and non-experts to display one's self to one's peers as the speaker displayed his own scientific credentials.

Registration for the Association meeting was, effectively, admission to certain civic and scientific spaces. As early as the Edinburgh 1834 meeting, tickets to Association meetings were also maps of the meeting's locations: Bristol's Local Committee in 1836 even instructed local printers that the Bristol ticket map should be no larger than the Edinburgh one, for fear of causing offence.[36] As meetings grew – in size of audience and in numbers of participants and different settings – they tended to outstrip the capacity of the host city to house them in one place. As

Geographies of civic science

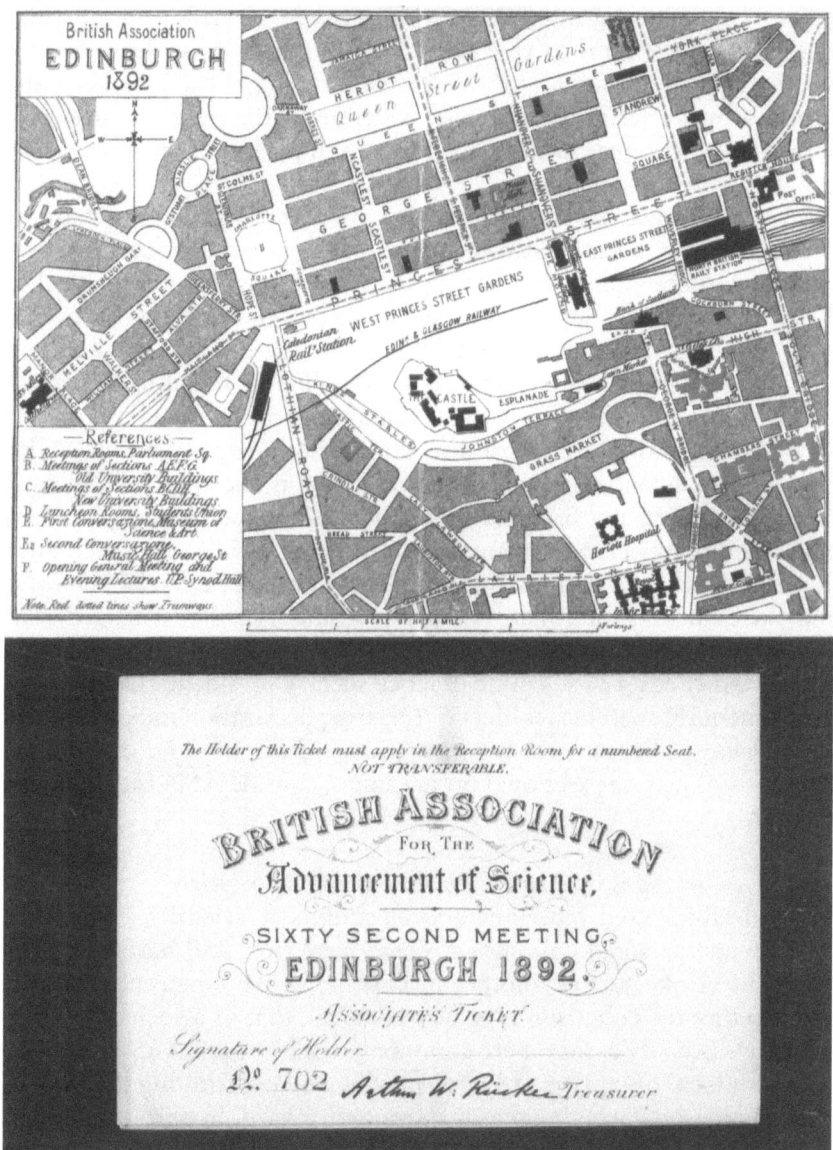

Figure 2.3 'Ticket map' and admission card for the BAAS meeting, Edinburgh, 1892

the ticket map for the Edinburgh meeting in 1892 suggests, attendance required careful planning of one's mobility within the city (Figure 2.3).

It is helpful, therefore, in examining how the Association worked and with what effect, to consider BAAS meetings not as one event, but in terms of their having multiple features – of lecture halls and museums as sites of speech and of viewing, of local collections which delegates would view as indicative of the scientific standing of the local area and as part of wider scientific and social wider networks – and to think, too, of the movement of audiences to different venues and of the intrinsically mobile nature of excursions.

BAAS meetings were not closed scientific worlds. Yet it is hard to know just how socially stratified they really were, among audiences especially. A commonplace of Victorian science was that many so-called 'local' or 'amateur' scientists were members of more than one body (and often of national and international bodies such as the BAAS) with interests and capacities in more than one discipline or scientific practice – such as taxonomy, field collecting, exhibiting, writing or public speaking – and, often, with positions of institutional authority in their discipline and municipal-civic responsibility in the town in question.[37] Looking at the different forms of BAAS science at work illuminates how particular places and practices were mutually constitutive settings for the conduct of science. But such illumination should not neglect the wider networks and issues with which many of the practitioners were engaged. The examples of museums as sites of display and excursions illustrate these points. Other cases of scientific practice such as presidential addresses – simultaneously speech sites, acts of scientific declaration and, as we will see, of geography, expressions of epistemic definition and disciplinary doubt – are, with the sectional programmes, considered in the following chapter.

Sites of display: locality and civic prestige

Most meetings used museums and existing civic scientific collections, even others' personal collections. There were over 250 natural history museums of one sort or another in England in the nineteenth century, and whether the collection type was personal, that of a society, municipal or a university's, they were commonly used by the BAAS in negotiation with local organisers.[38] Yet displaying and constituting the local as part of a national meeting was not straightforward. In part, it involved, from 1874, the production of handbooks and associated textual guides to the venues and region in which the meeting was set. BAAS handbooks provided a textual expression on thematic lines of the Association's provincial mission. Ordered for the most part chronologically, they offered in summary a review of the location's history, its local geography and scientific features: beginning with the local topography or geology, chapters covered natural history or more specialist subjects before covering industrial productions, notable sites and the scientific significance of the venue or the region. They did not, however, use local dialect to do so: in that sense, handbooks were about the international language of science, not the particular populist 'voice' of the towns or region in which they met: this is in contrast to the use in other contexts of the dialect 'voice' as one of the defining features Joyce identified in his discussion of the 'languages' of class and of social identity in later nineteenth-century and early twentieth-century England.[39] Most handbooks were written and collated by local figures, with members of local scientific bodies involved in the work: for this reason, BAAS handbooks may be understood as a textual expression of local science and a statement of local scientific capacity designed also to appeal more widely.

In text and in cabinet, scientific display required civic organisation of the appropriate specimens being used to constitute 'the local'. In Birmingham in 1886, for instance, the Natural History Sub-Committee of the Local Executive Committee established to organise the BAAS meeting that year reported upon initial difficulties in getting hold of local specimens for the intended exhibition:

> we issued a large number of circular letters inviting the owners of Natural History collections to lend specimens for exhibition. As it was desired to limit the exhibition to the locality of Birmingham and neighbourhood we have experienced some difficulty in obtaining specimens, but we are pleased to report that we have had promised various loans of collections which will ensure a good exhibit of the Fauna, Flora and Geology of Birmingham and the district.[40]

As inert specimens for display came in, so also the Mayor of Sutton Coldfield consented 'to the removal of living plants from Sutton Park to augment the illustration of the flora of the district', and members of the Birmingham Natural History and Microscopical Society organised a display to be held upon the evening of the meeting's *conversazione*.[41]

Display was not just about putting nature's artefacts on view. Displaying local artefacts, whether scientific or industrial, instilled pride in the hosts and allowed local knowledge to become national, for the duration of the Association meeting at least. In Bristol in 1898, the Bristol Museum authorities 'put themselves out of the way to do everything which can be done to make their valuable collection useful to the visitors. There will be an important series of local geological exhibits.'[42] If displaying depended in part upon local collecting – either by being in the field, or in the handbook, or by the more prosaic means of accumulating others' extant specimens – displaying and collecting were also sources of civic and scientific activity that benefited certain sciences more than others. Botanical work depended in at least one respect upon collaborative fieldwork and the safe return from the field of specimens then discussed in lectures and in private conversations. Botanical displays in BAAS meetings, by contrast, often focused on the microscopic examination of plants, upon mosses and lichens or, even, fossil specimens, rather than upon living plants which, in any case, featured when botanical gardens were open as a civic resort to BAAS delegates. Geological work readily lent itself to display, both of the inert specimens or of the products associated with geological work, such as maps and photographs. In these respects, the activities in botany and in geology in BAAS meetings mirror those practices of the many natural history societies of the time, bodies with which, as corresponding societies, the BAAS was anyway linked[43] (see Figure 2.2).

In relation to the BAAS Leeds meeting in 1890, members of the Leeds Geological Association reflected less upon the overall worth of the meeting that year (*cf.* the perceived collective benefits expressed of the 1858 meeting) than upon the advantages for their own science:

> The special feature of the present year has been the visit of the British Association to Leeds. Much was hoped for from the stimulus which, it was expected, would be given to scientific pursuits by the presence in our town of many eminent scientists. Though there has been no great accession of activity in our own Association as the result of these meetings, the Council feel that the prominence given to Yorkshire Geological work, especially in the Boulder and Photographic departments, ought to encourage the members to a determined endeavour for the attainment of a still higher standard of work in the future.[44]

Such evidence, whilst informative in one specific context, highlights a wider sense evident from assessment of BAAS meetings, namely that what was on display was not just the town but the region, the setting in which the meeting was held. The handbooks helped in this regard, and display cases spoke not just of interesting items but of an area that was scientifically engaging and of the inhabitants able to comprehend and interpret it – all reasons, after all, why towns appealed to the BAAS to hold their meeting there and not somewhere else. In turn, programmes for different sections reflected the economic or political interests of given towns where they could. Presidential addresses in later meetings would often highlight the scientific and industrial changes in the town in question since the last BAAS meeting there (as was the case noted above for Newcastle between 1838 and 1863). In Bristol in 1898, it was a source of regret that there were no papers on local geography: 'the geographical interest of the meeting was thus quite apart from the place, and had no reference to local conditions'.[45] In Newcastle in 1889, in

Hull in 1922 and in Liverpool in 1923, by contrast, the programme for section E geography was planned to reflect the commercial and imperial geographies of importance to those towns. 'In a commercial centre like Newcastle,' it was noted, 'we may fairly assume that the practical applications of geographical knowledge will receive prominent treatment by this section': attention was paid to trade routes and to commercial geography as well as to geographical education. In Hull, 'special attention was given to various aspects of the North Sea, geographical, geological, and biological, out of compliment to the town's position as a port'. In Liverpool, the city as a centre for Britain's imperial geography was the subject of several papers 'in view of the location of this year's meeting'.[46]

In this particular respect, the meetings of the BAAS helped promote the role of provincial geographical societies in these and other similar towns.[47] More generally, local scientific bodies in the region or town provided officers for the excursions and authors for the handbooks and so helped secure legitimacy for themselves and their members by association with the BAAS. Of course, programmes incorporated papers from visiting speakers, and local figures addressed topics of wider significance and used local examples to illustrate more general principles. There is, nevertheless, a sense that what was on display was the scientific capacity of local practitioners, the 'worth' of the area as much as the labelled specimens in cases and cabinets. 'Localness' was constituted through science and the geography of the area and so found expression in the organisation of the meetings.

The intention of BAAS organisers and local committees to ensure local interest was a concern throughout the later nineteenth century especially and was revisited with some urgency in the early twentieth century during debates about the Association's structure and future. Plans in 1909–10 to reorganise the sections noted 'That more attention should be paid to the previous selection of subjects, with particular interests to the places of the meetings; and that discussions should similarly be more carefully arranged'[48] (see Chapter 7). To judge from paper titles, meeting programmes always contained some element or other of local work, often as a strategy to ensure that science in the host setting had, if not equal space with the concerns of visitors, a prominent place nevertheless. In the South African meetings of 1905 and 1929, for instance, the greater attention paid to local topics in the latter meeting reflected a deliberate move to 'South Africanise' science. In the 1914 Australia meeting, discussions were even held over what proportion of the meetings and for which sections 'local' topics were to figure, with a 'one third rule with extension to one half in Zoology, Geology, Botany, and Agriculture'.[49]

The issues of 'localness' and the display of scientific culture locally that engaged the BAAS were not about equating provincial science with parochialism but part of the workings of a foundational strategy with respect to different urban and scientific circumstances. The sort of science that could be engaged with and the local/extra-local balance that could be struck in its delivery depended upon the place in which the meeting was held. This local or provincial agenda for association science was apparent also in excursions. Excursions allowed scientific and social agenda to be realised in and through specific settings and in ways which encouraged information about those settings to move well beyond them: as conversations in the field, as publications in learned journals and, notably, as guides and handbooks which depicted the local area and made it travel in print.

Walking and talking science: excursions and civic culture
In considering excursions as elements within the historical geographies of BAAS science, initial distinction might be made between inner-urban trips in which sites within the city illustrated matters of scientific interest, and sites outwith the immediate urban setting. The Manchester meeting of 1861, for example, was notable for its use of local industrial sites to illustrate the connections between science and technology, industry and commerce: chemical works, copper mines, the coal mines at Astley deep pits, Manchester's waterworks and so on.[50] This was a common feature of later BAAS meetings and of the sections on economics, statistics and engineering in particular, and for some larger towns, guidebooks and maps were even prepared to aid discovery of those venues deemed 'scientific' for BAAS audiences. On occasion, scientific excursionists themselves became objects of scrutiny: the *Times* reported how Nottinghamshire coal miners and their families were amused at 'the inartistic way' in which 'men with great names in science' behaved as they struggled with miners' lamps and equipment during a visit to a pit in 1866.[51]

Distinctions based solely on location within or outwith the city and solely in terms of particular sectional or scientific interests are hard to sustain. Excursions and site visits fulfilled a social as well as an intellectual function, bringing together locals with visitors, 'experts' with 'amateurs'. Moreover, recovering the intentions of those who visited exhibitions or made trips in the field is difficult. For some, science in the field provided a justification for sociability and was itself not the primary concern. During meetings, one could walk the city as a differentiated site for science's display but without having science in mind. Take Lady Caroline Howard, for example, in Dublin for the 1857 BAAS meeting. For her, attendance at morning paper sessions in geology, geography and ethnology, at the display of geological maps, the *conversazione* and afternoon promenade in the Zoological Gardens and at an evening's *soirée* at the Royal Irish Academy was a chance to converse with friends, to see and to be seen – 'I saw Judge Crampton and Lord and Lady Meath and several faces I know' – and to be amused rather than educated by science.[52]

Excursions were social and scientific affairs. At the 1878 Dublin meeting, members taking part in the geological excursion to Kilruddery (over 100 strong) were entertained to dinner by the Earl and Countess of Meath. The excursion allowed for enthusiastic amateurs to undertake field work and hold discussions with experts and provided for others an opportunity to converse or botanise as they walked:

> Scarcely had the long line of excursionists moved up the road than hammers were quickly displayed, and ladies and gentlemen, old and young, were seen most amusingly to the non-scientific observer peering into crevices of rocks, breaking off pieces of stone, and holding consultations as to what formation they belonged. Eagerly bent was seen many a fair scientist, rapping with her hammer at the rocks, and examining through her spectacles, for it must be confessed that some of these geologically-ladies [*sic*] wore glasses of studious import, and indicative of midnight oil expenditure, the fragments wore off. Some, however, who evidently more enjoyed the 'outing' than they were desirous of obtaining information about the Cambrian formation, looked on, strove to look learned, and sighed, others devoted their attention to the flora of the district.[53]

Excursions became a more notable feature of BAAS meetings during the second half of the nineteenth century. The overseas meetings were even likened to gigantic excursions, with ocean travel affording various opportunities for science on the move, and continental rail travel providing a means to observe the colonies and their peoples during and after the paper sessions (issues discussed in Chapter 4). Within Britain by the later nineteenth century, excursions were increasingly tailored to accommodate different specialist and general interests. Expansions in the rail network helped extend science. In Edinburgh, for the 1892 meeting, scientific site visits were planned within the city, within the local area and throughout central Scotland. Given the excellence of the rail network, the excursions to Tayside, Argyll and to the Ben Nevis Meteorological Observatory started at 04.00 a.m. and brought delegates back to the capital by late evening: Britain's highest mountain and, in the case of the Observatory, Britain's highest and most isolated scientific recording laboratory became public scientific sites and sights for a day. One report of the 1901 Glasgow meeting speaks to matters of practice and social and intellectual intention more generally held: 'Excursions will be arranged by which Members will be afforded an opportunity of visiting locations unsurpassed for the beauty of their natural scenery; presenting special attractions for the scientific explorer, whether Geologist, Zoologist, or botanist; for the Archaeologist, and those who desire to observe the various important Industries of the district of which Glasgow is the centre.'[54] In short, excursions – even as they required certain forms of dress and comportment in the town or country – ensured a form of regional survey and subject specific study (Figure 2.4).

Beyond the urban setting, a local-cum-regional geographical context was important to the success of BAAS meetings. Particular sites of scientific interest were the subject of attention, sites were tailored to given sectional interests and were evaluated in relation to the demands of associated social activities (overnight accommodation, dinner, even entertainment at the expense of local nobility or gentry if such were BAAS members). We should be careful, therefore, about proposing any strict classification or typology for BAAS scientific excursions and careful too in thinking about the historical geographies of 'in-the-field' science in terms of any strict separation between issues of production, reception and mobility. What may matter more are the connections in practice between given sites and spaces and scientific audiences. BAAS meetings were not city-wide affairs but were, rather, mobile matters of practice in specific socio-scientific settings in and out of the city for audiences whose intentions and capacities were not the same (Figure 2.5).

Experiencing science locally: questions of reception

Identifying the reception of BAAS science

BAAS papers are not systematic in what they record about public reaction to meetings. Attendance figures were not consistently kept and do not permit us to know who in any given total went to what sectional programme or other event[55] (Table 2.1). To generalise from attendance figures between 1831 and 1931: attendance was generally higher in the northern towns; women made up about one-third of the delegates, certainly from the mid 1860s, although women as a fraction of the audience seemed to decline somewhat from the later 1880s; attendance was highest in Manchester for the 1887 (save for the rather unusual circumstances of the 1914 Australia meeting and the 1931 London centenary meeting); and most meetings had only handfuls of foreign delegates, except where the port location of the meeting – Ipswich in 1851, Norwich in 1868, Dundee in 1912, Liverpool in 1923 – allowed ease of access and so heightened their numbers.

Geographies of civic science

Figure 2.4 Depictions of excursion science from the 1865 Birmingham BAAS meeting. Upper: 'The Wrekin.' Lower: 'Through the woods to the Wrekin – "the pursuit of science under difficulties"'

Figure 2.5 Depictions of science as performance, display and sociability, from the BAAS meeting, Cambridge, 1845: lecture demonstrations by Professor Brontigny and, to the foot, a 'floralia' in the gardens of Downing College

Table 2.1 Attendance at BAAS annual meetings, 1831–1931

Location	Total No. in attendance	Ladies		Foreigners (No.)
		No.	As % of overall attendance	
York 1831	353	–	–	–
Oxford 1832	–	–	–	–
Cambridge 1833	900	–	–	–
Edinburgh 1834	1,298	–	–	–
Dublin 1835	–	–	–	–
Bristol 1836	1,350	–	–	–
Liverpool 1837	1,840	–	–	–
Newcastle 1838	2,400	1,100	45	–
Birmingham 1839	1,438	–	–	34
Glasgow 1840	1,353	–	–	40
Plymouth 1841	891	60	7	–
Manchester 1842	1,315	331	25	28
Cork 1843	–	160	–	–
York 1844	–	260	–	–
Cambridge 1845	1,079	172	16	35
Southampton 1846	857	196	23	36
Oxford 1847	1,320	203	15	53
Swansea 1848	819	197	24	15
Birmingham 1849	1,071	237	22	22
Edinburgh 1850	1,241	273	22	44
Ipswich 1851	710	141	20	37
Belfast 1852	1,108	292	26	9
Hull 1853	876	236	27	6
Liverpool 1854	1,802	524	29	10
Glasgow 1855	2,133	543	25	26
Cheltenham 1856	1,115	346	31	9
Dublin 1857	2,022	569	28	26
Leeds 1858	1,698	509	30	13
Aberdeen 1859	2,564	821	32	22
Oxford 1860	1,689	463	27	3
Manchester 1861	3,138	791	25	15
Cambridge 1862	1,161	242	21	25
Newcastle 1863	3,335	1,004	30	25
Bath 1864	2,802	1,058	38	13
Birmingham 1865	1,997	508	25	23
Nottingham 1866	2,303	771	33	11
Dundee 1867	2,444	771	31	7
Norwich 1868	2,004	682	34	45

Table 2.1 (continued)

Location	Total No. in attendance	Ladies No.	Ladies As % of overall attendance	Foreigners (No.)
Exeter 1869	1,856	600	32	17
Liverpool 1870	2,878	910	31	14
Edinburgh 1871	2,463	754	30	21
Brighton 1872	2,533	912	36	43
Bradford 1873	1,983	601	30	11
Belfast 1874	1,951	630	33	12
Bristol 1875	2,248	672	30	17
Glasgow 1876	2,774	712	26	25
Plymouth 1877	1,229	283	23	11
Dublin 1878	2,578	674	26	17
Sheffield 1879	1,404	349	25	13
Swansea 1880	915	147	16	12
York 1881	2,557	514	20	24
Southampton 1882	1,253	189	15	21
Southport 1883	2,714	841	31	5
Montreal 1884	1,777	74	4	86
Aberdeen 1885	2,203	447	20	6
Birmingham 1886	2,453	429	17	11
Manchester 1887	3,838	493	13	92
Bath 1888	1,984	506	26	12
Newcastle 1889	2,437	579	24	21
Leeds 1890	1,775	334	19	12
Cardiff 1891	1,497	107	7	35
Edinburgh 1892	2,070	439	21	50
Nottingham 1893	1,661	268	16	17
Oxford 1894	2,321	451	19	77
Ipswich 1895	1,324	261	20	22
Liverpool 1896	3,181	873	27	41
Toronto 1897	1,362	100	7	41
Bristol 1898	2,446	639	26	33
Dover 1899	1,403	120	9	27
Bradford 1900	1,915	482	25	9
Glasgow 1901	1,912	246	13	20
Belfast 1902	1,620	305	19	6
Southport 1903	1,754	365	21	21
Cambridge 1904	2,789	317	11	121
South Africa 1905	2,130	181	9	16
York 1906	1,972	352	18	22

Geographies of civic science

Table 2.1 (continued)

Location	Total No. in attendance	Ladies No.	Ladies As % of overall attendance	Foreigners (No.)
Leicester 1907	1,647	251	15	42
Dublin 1908	2,297	222	10	14
Winnipeg 1909	1,468	90	6	7
Sheffield 1910	1,449	123	8	8
Portsmouth 1911	1,241	81	7	31
Dundee 1912	2,504	359	14	88
Birmingham 1913	2,643	291	11	20
Australia 1914	5,044	–	–	21
Manchester 1915	1,441	141	10	8
Newcastle 1916	826	73	9	–
[No meeting 1917]				
[No meeting 1918]				
Bournemouth 1919	1,482	153	10	3
Cardiff 1920	1,380	–	–	20
Edinburgh 1921	2,768	–	–	22
Hull 1922	1,730	–	–	24
Liverpool 1923	3,296	–	–	308
Toronto 1924	2,818	–	–	139
Southampton 1925	1,782	–	–	74
Oxford 1926	3,722	–	–	69
Leeds 1927	2,670	–	–	161
Glasgow 1928	3,074	–	–	74
South Africa 1929	1,754	–	–	83
Bristol 1930	2,639	–	–	54
London 1931	5,702	–	–	449

Source: *Report of the Centenary Meeting of the British Association for the Advancement of Science* (London: British Association, 1931), pp. xii–xv.

It is harder still to know what delegates thought of the meetings. Diaries permit insight into individuals' engagement with the activities of the BAAS, but we must beware the dangers of generalisation from such sources.[56] Professional journals often covered BAAS meetings for given sciences – as we shall see of geography – but here, too, we must recognise the possibilities that expressed professional differences masked personal ones and that reportage was constrained by editorial and publishers' conventions.

Newspaper and other sources can reveal matters that BAAS officers might not have welcomed being made public. Even as towns and cities generally welcomed the BAAS and planned for its meeting in different ways and civic spaces, the nature of the urban setting could determine where, how, and to a degree even if Association science was received by the public. In Glasgow in 1901, for instance, the geography programme was not well attended as a result of the venue for that section's paper sessions:

> Although the University was fixed upon as the general headquarters of the Meeting, yet, in consequence of an apprehension that the space available at the University was insufficient, it was arranged that Section E (Geography) should meet at the Queen's Rooms [a civic hall distant from the university]. It had always been the aim and desire of the Executive to keep the whole sections together; and it is a matter of regret that the separation of the Geographical from the other sections had the effect of restricting the attendance upon one of the most interesting sections of the meeting.[57]

A quarter of a century earlier, public reaction to the 1876 meeting, distilled in numerous Glasgow newspapers, was strongly of the view that the whole affair had been a failure. For one commentator, this was due to incorrect prior perceptions about the nature of the BAAS and of its meetings: 'A good deal of this is unquestionably due to the disappointment of a large number who tacitly assume that gatherings of the kind, breaking in upon the monotony of business, ought to be amusing. The coming of the Association was looked forward to as a kind of entertainment, and those who took that view of its functions have been considerably annoyed by the seriousness of its actual behaviour.' Behind this observation lay a deeper disquiet with the BAAS:

> The Association is trying to serve two masters – science and the public. If it sticks to science its meeting becomes superfluous, or ought to be limited to scientific men. If it seeks to serve the public in a way the public can appreciate it must meet once in ten years, and have something to show that the unscientific mind may grasp and feel interest in. It may, indeed, be intended to show what charming people men of science are, and to induce the public to adopt their pursuits from desire to acquire their fascinations. But in that case the science may be dispensed with altogether, and the lighter graces of the man of science in his domestic and social aspects cultivated in its stead. The title would then read – British Association for Popularising the Scientific Men.[58]

Although initial disquiet about the purpose, content and future of the BAAS had largely ceased by 1845, public doubts like this were aired quite commonly, and the Association, notably because of its annual meeting, the oddities of 'the gentlemen of science' and the strange pronunciations of science's terminology, was regularly satirised in *Punch* (Figure 2.6). Where, earlier, the BAAS had been mocked as 'The British Association for the Advancement of Everything in General and Nothing in Particular' – or referred to as a 'band of itinerant twaddlers'[59] – later comments tended to poke fun more at the characteristics of the inhabitants of the host town or at notable goings-on in the scientific proceedings: the explorers David Livingstone and Henry Morton Stanley, for example, were satirised in coverage of the 1872 Brighton meeting, given the latter's 'discovery' of the former in Africa[60] (Figure 2.7).

Political topographies of reception: the Dublin meetings
In Ireland, the BAAS was received with animosity at several of its meetings not because of the science but because it was British: the meeting, and its week-long events of urbane scientific sociability, was associated with political ascendancy. This is a consistent theme in reception of the Dublin meetings, certainly in the later meetings there. Isaac Butt, writing in the *Dublin University Magazine* in respect of the 1835 meeting, predicted the rapid demise of a body too taken up with display, not enough with science: 'We cannot conceal our conviction that for the purposes of the advancement of science the association is little better than useless. It gives scientific men, or men who call themselves scientific, a week of pleasuring at the expense of three weeks' idleness, but it does nothing more.'[61] Butt was an advocate of a 'progressive Tory Protestantism', keen on Irish independence but not on separation from the British Empire. His federalist position was rooted in a belief that there was a sufficiency and strength to Irish intellectual life in contrast to what he took to be the impermanent and shallow work of the British Association. At a time of unrest, given agrarian protest in south-east Ireland against absentee landlordism and a rising sentiment against Protestant rule, the local organisers of the 1835 meeting nevertheless banned certain sections of the press from attendance and from reporting on it. This did not prevent the Catholic and nationalist *Freeman's Journal and Daily Commercial Advertiser* from recording how 'the managers of the affair in Dublin, who are a clique of bitter Orangeists', manipulated the reception of the meeting by ensuring coverage was carried only in Protestant papers: 'the Orange newspapers, which are patronised by the committee of mismanagement, will give ample details, as they have hitherto been alone favoured with the advertisements'. The 'miserable, consumptive Orange clique' was in fact the College of Physicians: by contrast, the College of Surgeons offered 'no niggardliness, no paltriness, no exclusiveness' in hosting a scientific breakfast for leading Irishmen of science, whatever their religious and political leanings.[62]

Figure 2.6 Problems of comprehension, gently satirised in *Punch*. 'Things one would rather have left unsaid. "Ach! Gracious laty, I hope zat my long Cherman Lecture on ze Boetical Aspects of ze Bliocene Beriod did not *bore* you very much zis afternoon?" "Oh, not at all, Professor Wohlgemuth. I don't *understand* German, you know"'

Geographies of civic science 53

Figure 2.7 The geographer-explorer at rest. David Livingstone at rest over the Africa he did so much to discover. (In the accompanying text Livingstone is made to resent his own 'discovery' by H. M. Stanley)

Similar issues resurfaced in 1857. To the reporter of the *Dublin Evening Post* the presidential and opening address of the 1857 Dublin meeting was 'well calculated to impress our foreign visitors' (by which he meant British as well as other overseas visitors) 'with the high estimation in which science is held in this country, and the adaptability of the Irish mind to her practical culture'. Appeals to the Dublin citizenry for financial support for the meeting – and for them to accommodate, gratis, visiting scientists – had been made: 'We think a selfish sense of interest ought to induce the people of Dublin to subscribe to it at once.' Papers on Irish dialectology – which 'afforded not a little amusement to the audience' – and on the craniometry of Irish and Scottish Gaels were cited as an indication that the BAAS had planned at least elements of its sectional meetings with local audiences' needs in mind and so should be welcomed.[63]

By contrast, the strongly pro-independence newspaper *The Nation* used the visit of the Association to reiterate political arguments about domination by Britain and satirised the Association precisely because its delegates were dependent upon Dublin residents for their board and lodging:

> Among the savans now assembled in the metropolis of Ireland, there are many whose names are emblazoned on the golden roll of Philosophy, Science and Art. Such men ought to be welcome in Ireland: here they tread a land which was once the home of learning, the munificent patroness of the Arts and Sciences, ere the country for which the Association takes its name, had emerged from the night of barbaric ignorance. Here they will find traces of all that interest the ethnologist, and the antiquarian; they will find relics of a glorious past, evidence of a miserable present. They may employ themselves profitably in investigating the cause of this state of things; in tracing the date at which this decadence set in, and they will find that when Ireland ceased to be independent, the Arts and Sciences fled the land.

The report continued, in the form of a BAAS presidential address to the 'Dublin Provisional [i.e. republican] Committee of the British Association', to mock the needs of visiting scientists to have any need for food and shelter:

> It strikes me that men of such notoriously studious habits and concentrated scientific ambition, require very little sleep at all (cheers). Besides they do not visit us for the purpose of sleeping or enjoying themselves in any way, but for that of instructing us. As to their diet, I would fain think they are amply provided for. Ethnography shows us that Newton (who required, by the way, only four hours' sleep) often forgot his dinner for two consecutive days (loud cries of hear, hear). A single water melon sufficed Galileo for a similar period (bursts of applause); and I rejoice to have to read that La Place, while calculating the attractive force which the Star X in the Dumb Bell Nebulae, exercised upon our Moon, pursued his mathematical studies so far as to destroy his appetite altogether. Facts like these, gentlemen, lead me to hope that, in respect to the preparations made for the British Association, the citizens of Dublin have nothing to fear. To a mind that would square the circle, what possible interest could attach to an article so insignificant as a mutton cutlet?[64]

Half a century later, the advertisement which hung outside Trinity College was defaced – 'maliciously damaged to read "THE BRITISH Ass"', as the *Irish Times* had it. Within the meeting, the fraternity of science was perceived to overcome factional political differences: 'It is not the smallest of our debts to the British Association that it provided us with a platform on which Irishmen of all parties have been able to exchange views in the true spirit of scientific enquiry.'[65]

Women as audience in the BAAS

At the Association's foundation in 1831 women were formally excluded from BAAS paper sessions but encouraged to accompany their scientific husbands and relations in order to add social status and glamour to the meetings. Even when barred, however, women attended sectional meetings and, from 1838, were formally admitted into all sections except section D, which then included zoology, a subject deemed unsuitable, given its attention to matters of biological reproduction. By 1839, however, women could attend all sectional sessions but only in certain parts of the sites for science – 'the galleries only or railed-off spaces' – where that was possible in the civic buildings in question and could be enforced.[66] Women were not allowed to become full members of the Association until at least the 1850s and were at first sold special Ladies' Tickets (which ceased to be separately issued from 1919). Morrell and Thackray have argued that the presence of large numbers of women quickly became essential because their financial contribution was irreplaceable and that, because of their presence in the social elements of meetings, they acted as social 'cement'.[67] Women attending BAAS meetings were typically related to visiting men of science or were members of the host town's leisured classes. As the century progressed, however, increasing numbers of women paid their £1 as a result of their own interest in the topics.

Whether or not Lady Caroline Howard read the reports of 1857 is unknown. She was one of 569 women delegates at the BAAS meeting that year (Table 2.1). Her engagement with the BAAS in Dublin was as we have seen about sociability rather more than science, and her politics (we may presume) contrasted with those expressed in *The Nation*. Lady Caroline attended one set of geography papers 'but the room was greatly crowded and so we did not hear much'. Her companions were even less fortunate in attending Livingstone's lecture on African discoveries: 'Julie enjoyed herself so much and brought me back such an account of it that I felt quite in despair at being laid up. They however did not hear one word of the lecture as they got bad seats, and Dr Livingstone speaks in a whisper.'[68] We may never fully know why an inaudible lecture on African exploration should have been so enjoyed, but what surely attracted them was Livingstone's celebrity status, much less his geography.

Such illustrations point to the complexities surrounding the 'reception' of BAAS science. For some, the Association's coming was political anathema and of national importance. For others, it was an opportunity for polite education, to attend and to observe and to listen, but not always to hear. Lady Howard and her companions were far from unusual delegates to the BAAS meetings. Although we can never know in detail who was at given meetings and why, it is clear that women were present in large number at BAAS meetings (see Table 2.1) and that women attended some sciences more readily than others. For such reasons, we must again be careful in seeing BAAS meetings as single events, their audiences for science as homogeneous.

For some, like Lady Caroline in Dublin in 1857, women were attracted by celebrity: large numbers of women went at one time or another to hear Sir John Ross, Michael Faraday, Sir John Herschel and Sir David Brewster, for example, often examining the faces and manners of those men made famous by lecture tours, books and newspaper accounts of their scientific endeavour to see what a distinguished scientist looked and sounded like. But where women went in numbers to botany, entomology and geology and to geography, section A (mathematics and physical science) was reckoned hard and masculine. Thus even when *The Times* in 1853 noted 'a growing desire among the lady inhabitants of the towns successively visited to avail themselves of the opportunities for scientific information afforded them by the presence of the *savans*', reporters still highlighted the difference between these attendees and the scientific members: 'the lady subscribers have sometimes neglected the mathematical and physical section and have overlooked the poetry of statistics'.[69]

The diary of Agnes Hudson, who attended the 1875 Bristol and 1879 Sheffield meetings, illuminates these issues. On the evening of 25 August 1875, she and her companion set out 'to attend the first general meeting'. 'When we reached the Hall which is a handsome, but insufficiently ventilated building we found the heat intolerable and we could not hear a word that Sir John Hawkshaw, the President said we soon withdrew, determing [sic] to read his address next day.' (Official addresses being printed in the following day's newspapers.) The next day:

> We went to . . . Section D. on Anthropology. We sat there a long while doing nothing, for the papers were not yet begun to be read till an hour after the right time, it was rather amusing when the proceedings did begin for a gentleman in the audience had to rise and defend himself against some accusations made by the speaker, Mr. Pengelly, and they got quite excited over it . . . . When we had had enough of the Section we went to the Victoria Rooms and had lunch . . . In the afternoon we walked up Brandon Hill which is quite close and whence you have a very good view of the whole of Bristol, on our way home we ordered some flowers for our hair and then dressed for dinner for the first time since our arrival here. At 8 p.m. we started in a cab for the Colston Hall where was held the microscopical Soirée, we encountered Mr Hart there who was much surprised to

find that we had no one to look after us, and who informed us that some
of the objects under the microscopes were well worth looking at, we found
however that it required more patience than we possessed, the hall was
crowded and very hot, which necessitated our imbibing a large quantity of
ices, which were remarkably good for such an occasion. We arrived home
safely at 10.30.

Next morning, Hudson recorded, the anthropological section sessions were so crowded that 'several persons sat on the mantelpiece'. Later in the week (31 August):

we went to the geographical Section where a paper was read on the Arctic
expedition by Mr Markham [Clements Markham: president of section E
in Sheffield in 1879] who had just returned to England on the *Valorous*,
Colonel Montgomerie read a paper on the Himalayas, he pronounced it
Himâlia, which was not at all interesting and which I do not think anyone
would have listened to but that the Arctic expedition came next.[70]

Four years later, in Sheffield, she was again disappointed by section E proceedings: 'Major Serpa Pinto [a leading Portuguese explorer] was to read a paper on his travels in Africa, but he deputed an English man to read it for him, and we did not find it so interesting as we had expected.' An afternoon excursion to steel mills followed: 'We returned to Sheffield at 8.20, and went home by ourselves, many of our fellow travellers hurried off to an evening lecture on radiant matter, but we went straight home.'[71]

Whilst women were in general depicted as unscientific and the subjects whose business they more commonly attended were, in part, often seen as such by association, the BAAS nevertheless came to depend upon the presence of women for reasons other than the financial and the decorous. By feminising Association audiences, women succeeded in masculinising the scientific expertise of the male practitioners, lending the gentlemen of science additional authority. Women, in short, helped define science's public as something opposed to its practitioners, especially so in geography, as we shall see.[72]

Conclusion

This chapter has shown how BAAS meetings – the principal expression of its stated provincialising geographical mission – worked locally and nationally. In illustrating the importance of the Association's urban settings, there are parallels with that work which has pointed to the urban history of science as a history of scientific sites in action, whether as expertise, science's representation in the city, places of knowledge in urban context, or knowledge from the street.[73] The BAAS clearly drew upon local urban and scientific expertise: scientific institutions whose members variously formed civic delegations to invite the Association, or who led BAAS excursions, and who wrote and edited handbooks. Science was seen in terms of civic benefit: benefit to the Association in seeing local sites of wider relevance, benefit for the host town in having experts visit. Museums, meeting rooms, botanical gardens and other civic spaces were places of scientific knowledge and of social display even if, during BAAS meetings, they were only temporary venues for engaging in science, ticketed sites in which one had but transient engagement with science and with scientists (Figure 2.8). So, too, BAAS handbooks and excursion guides explored and represented the city and the region, their authors and field leaders interpreting sites in ways which might equate with knowledge from, or perhaps of, the street. In that regard, we may concur with the view that 'when the city is involved, the historian of science must pay as close attention to it as to the science conducted there'.[74]

In contributing to an historical geography of science rather than an urban history of science in which specialist institutions of science were 'fixed' within cities, this examination of the BAAS has disclosed different spaces and places – different settings – in which science was temporarily housed and it has revealed the importance of particular sites beyond the city. One consequence is to emphasise the importance of different scales of geographical analysis. The BAAS had an international dimension notably after 1884 and its overseas meetings but also in terms of the individual foreign scientists attending its meetings. What was a national body depended for its success upon geographical mobility within the nation, upon a provincial or regional agenda that was a foundational policy and a consequence of science being at work locally in the host towns. Within cities, certain civic spaces were opened up for the consumption and display of science as were locales of significance near by. As Livingstone has remarked, 'Precisely what the correct scale of analysis is at which to conduct any particular enquiry into the historical geography of science – site, region, nation, globe – has to be faced.'[75] Geographical study of the BAAS meetings has here documented less the 'correct' scale of analysis and more the importance of reading across scales to see connections: a national policy of provincialism, city-wide work, local sites and venues and practices of 'place making' and of science's making such as illustrated talks and the handbook which depended upon and promoted regional survey, field sites, lecture halls and display rooms.

A further consequence of illuminating the different civic spaces, places

Geographies of civic science

Figure 2.8 'Tableaux-Vivants from the History of Science.' Part of the theatrical display of moments of scientific importance at the BAAS meeting, Edinburgh, 1892. (That for geography portrayed Columbus's discovery of America in 1492)

and scales in which science was at work in BAAS meetings in Britain is to challenge the city or the town as the necessary unit of analysis in any historical geography of science. Of course, there will still be benefits – sources permitting – of looking over time at BAAS meetings in the same place. Many meetings – Newcastle in 1863, Edinburgh in 1871 and 1892, Glasgow in 1901, Leeds in 1927 – were chosen because of successful earlier meetings there, not just because the Association considered it necessary to return to such 'provincial' locations. It is also clear that the city was not a setting in any simple locational or 'containing' sense but, rather, a network of civic venues, of urban 'field sites' and a locus from which regional survey work was undertaken. Different practices in different settings – sitting attentively, conversing, hammering at rock exposures, reading handbooks, walking past display cases or promenading to be seen – were each part of the business of Association science.

This consideration of the ways in which the BAAS meetings worked and how they were received has identified something of the ways in which an urban historical geography of science transcends 'local' geographical matters whilst being attentive to the social practices through which science was consumed, received and circulated. If, then, science is not one thing, BAAS meetings were differentiated and the city is no longer becomes the required unit of assessment, our focus can more readily become certain sites, matters of mobility and different social intentions and scientific practises. Other topics include audiences' participation by social rank or by scientific affiliation, differences in perception and purpose and, potentially, of variation in attendance in relation to scientific subject. Attending did not necessarily mean understanding (or even hearing) what was said. Being at a BAAS meeting as a delegate was a form of scientific production *and* reception. Being at the meeting required timing one's day and location within and perhaps beyond the city – to consume civic science demanded mobility and engagement with city life and local/regional site.

Attention to questions of setting – understood not at a city-wide scale but as complex places in which different scientific practices and social interests came together – may help move us away from the dualism of sites/places of production and sites/places of reception. It is clear for many towns that the location of the BAAS meeting was chosen because of the prior existence of 'local science' or in response to the need to go where it had not before gone. There is evidence to suggest that the prospective arrival of the BAAS acted to constitute local science: that is, by being on a national circuit for Association meetings, local or provincial science and the social and intellectual networks that sustained it were either brought into being or given renewed vigour by virtue of being an intended Association venue. Consideration of BAAS meetings in terms of their different settings raises questions about the mobility and make-up of Association science given that the subjects that made up the BAAS sections were themselves in the process of formation.[76] Attention to the practices involved in the meetings also highlights the complexities involved in the making of science as 'provincial' given that visiting scientists were commonly perceived to bring 'expertise' with them even as 'local knowledge' was being constituted and enhanced by the fact of the BAAS meeting. With these ideas in mind, let me turn to the science and conduct of geography in the British Association.

Notes

1 MS. Dep. BAAS 5, Miscellaneous papers, 1831–1869, fol. 1. In what follows, all citations from the BAAS papers held in the Bodleian Library, Oxford (the principal BAAS archive) are given in this form.
2 J. Morrell and A. Thackray, *Gentlemen of Science: Early Years of the British Association for the Advancement of Science* (Oxford: Clarendon Press, 1981), p. 98.
3 Morrell and Thackray, *Gentlemen of Science*, pp. 104, 126–7.
4 Morrell and Thackray, *Gentlemen of Science*, p. 129.
5 Morrell and Thackray, *Gentlemen of Science*, p. 161.
6 R. M. MacLeod, J. R. Friday and C. Gregor, *The Corresponding Societies of the British Association for the Advancement of Science, 1883–1929* (London: Mansell, 1975).
7 Morrell and Thackray, *Gentlemen of Science*, pp. 164–222; P. Lowe, 'The British Association and the provincial public', in R. MacLeod and P. Collins (eds), *The Parliament of Science: The British Association for the Advancement of Science, 1831–1981* (Northwood: Science Reviews, 1981), pp. 118–44; A. D. Orange, 'The British Association for the Advancement of Science: the provincial background', *Science Studies* 1 (1971), pp. 315–29; J. M. Edmonds and R. A. Beardmore, 'John Phillips and the early meetings of the British Association', *The Advancement of Science*, 12 (1955), pp. 97–104; C. W. J. Withers, R. Higgitt and D. A. Finnegan, 'Historical geographies of provincial science: themes in the setting and reception of the British Association for the Advancement of Science in Britain and Ireland, 1831–c. 1939', *British Journal for the History of Science*, 41 (2008), pp. 385–415.
8 On the BAAS overseas, see M. Worboys, 'The British Association and empire: science and social imperialism', in MacLeod and Collins (eds), *Parliament of Science*, pp. 170–87; S. Dubow, 'A commonwealth of science: the British Association in South Africa, 1905 and 1929', in S. Dubow (ed.), *Science and Society in Southern Africa* (Manchester: Manchester University Press, 2000), pp. 66–99.

9 MS. Dep. BAAS 5, Miscellaneous papers, 1831–1869, fol. 18.
10 MS. Dep. BAAS 142, Correspondence relating to Annual General Meetings, fol. 96, Birmingham, 29 August 1839.
11 *Aris's Birmingham Gazette*, 2 September 1839.
12 MS. Dep. BAAS 18, Printed minutes of the Council Meetings, 1841–1857, fol. 49, London, 14 April 1847.
13 MS. Dep. BAAS 142, Correspondence relating to Annual General Meetings, fols 249–252, Oxford, 26 June 1847.
14 L. Miskell, 'The making of a new "Welsh metropolis": science, leisure and industry in early nineteenth-century Swansea', *History* 88 (2003), pp. 32–52, especially pp. 40–8.
15 MS. Dep. BAAS 142, Correspondence relating to Annual General Meetings, fol. 1, Ipswich, 2 August 1848.
16 MS. Dep. BAAS 18, Printed minutes of the Council meetings, fol. 49, London, 14 January 1848.
17 MS. Dep. BAAS 18, fol. 54, Swansea, 9 August 1848.
18 MS. Dep. BAAS 18, fol. 119, Dublin, 31 August 1857.
19 Leeds University Library, Leeds Philosophical and Literary Society, MS. 28 A, pp. 16–17.
20 Leeds University Library, Leeds Philosophical and Literary Society, MS. 28A, p. 3.
21 Lowe, 'The British Association and the provincial public', pp. 123–6. Lowe is correct to note that this was published once the 1853 Hull meeting had been confirmed – 'some months before it was due' – but neglects to point out that Frost initially delivered his views as a spoken address in advance of the meeting.
22 C. Frost, *On the Prospective Advantages of a Visit to the Town of Hull by the British Association for the Advancement of Science* (Hull: printed privately, 1853), p. 32.
23 Frost, *On the Prospective Advantages*, p. 26.
24 Committee Minutes of the Literary and Philosophical Society of Newcastle upon Tyne, 18 September 1862 [no pagination].
25 Correspondence Books of the Literary and Philosophical Society of Newcastle, 28 August 1862 [no pagination].
26 MS. Dep. BAAS 151 [no pagination].
27 [Anon.], 'Anthropology and the British Association', *Anthropological Review* 3 (1865), pp. 354–71, quote at p. 358.
28 MS. Dep. BAAS 409 [no pagination].
29 MS. Dep. BAAS 408 [no pagination].
30 MS. Dep. BAAS 12, Minutes of General Committee, 1869–1905, fol. 61, Glasgow, 11 September 1876.
31 MS. Dep. BAAS 12, fol. 84, Dublin, 19 August 1878.
32 Mitchell Library, MS. M.P.31 D-TC, 14.1.31, Glasgow Corporation Minutes relating to the Visit of the BAAS, 1901, fol. 518, 1 October 1896.

33 Cambridge University Archives, MS. CUR 111.2*, Letters 16 (from A. R. Forsyth, 9 November 1901) and 23 (from C. Heycock, 10 November 1901).
34 Lowe, 'The British Association and the provincial public', p. 135. More generally, no clear picture emerges from the later nineteenth century of the involvement of university communities in directing the BAAS to their towns, or of local authorities or scientific figures in towns seeking to use the visit of the BAAS to promote their own university or plans for one. Leading academics were certainly involved and university rooms often used and BAAS handbooks were often compiled by local university figures (see Chapter 7 for the example of Leeds in 1927) but no evidence has been identified to suggest, for example, that the visit of the BAAS to a non-university town was used to influence local demand for the establishment of a college or university there.
35 J. Secord, 'How scientific conversation became shop talk', *Transactions of the Royal Historical Society*, Sixth Series, XVII (2007), pp. 129–56.
36 Bristol Record Office, MS. 32079 (39), Minute Book of the Local Council for the Reception of the British Association, 1 October 1835–20 September 1836, p. 26, 16 May 1836.
37 This point is clear in works that discuss the nature of science and society in this period: see for example S. F. Cannon, *Science in Culture: The Early Victorian Period* (Cambridge MA: Harvard University Press: 1978); L. Goldman, *Science, Reform and Politics in Victorian Britain: The Social Science Association, 1857–1886* (Cambridge: Cambridge University Press, 2002); B. Lightman (ed.), *Victorian Science in Context* (Chicago: University of Chicago Press, 1997); R. MacLeod, *Public Science and Public Policy in Victorian England* (Aldershot: Variorum, 1996); F. M. Turner, *Contesting Cultural Authority: Essays in Victorian Intellectual Life* (Cambridge: Cambridge University Press, 1993).
38 S. J. M. M. Alberti, 'Placing nature: natural history collections and their owners in nineteenth-century provincial England', *British Journal for the History of Science*, 35 (2002), pp. 291–311.
39 P. Joyce, *Visions of the People: Industrial England and the Question of Class, 1848–1914* (Cambridge: Cambridge University Press, 1991), pp. 279–304.
40 Birmingham University Special Collections, MS. Ref: 4/i/3, Exhibition Sub-committee Minutes, 40, 11 August 1886.
41 Birmingham University Special Collections, MS. Ref: 4/i/3, fol. 43.
42 *Bristol Observer*, 27 August 1898.
43 D. A. Finnegan, *Natural History Societies and Civic Culture in Victorian Scotland* (London: Pickering & Chatto, 2009).
44 Leeds University Library, Leeds Geological Association Minutes, Dep/052, Box 5, Minute Book, Report of Council for the Session 1890–1891. The 'Boulder Department' refers to the fact that the BAAS encouraged local work on the identity and distribution of erratic rocks and boulders.
45 [Anon.], 'Geography at the British Association, Bristol, 1898', *Geographical Journal* 12 (1898), p. 377.

46 MS. Dep. BAAS 174, Newcastle Annual Meeting, p. 20. The Geography section there further noted that 'occasion may be taken to show what work has been done to open up markets for our goods'. The Hull quote is from the *Yorkshire Post*, 18 August 1923. On Liverpool, see MS. Dep. BAAS 425, fol. 54, and *Nature*, 1 September 1923.
47 J. M. MacKenzie, 'The provincial geographical societies in Britain, 1884–1914', in M. Bell, R. Butlin and M. Heffernan (eds), *Geography and Imperialism*, (Manchester: Manchester University Press, 1995), pp. 93–124.
48 MS. Dep. BAAS 30, Papers of the Committee of the Council on the Re-organisation of Sections, 1909–1911, fol. 61.
49 Adolph Basser Library, Australian Academy of Sciences, A. D. C. Rivett papers, MS. 83, Correspondence Files, MS. 83/26 3–6, O. Masson to A. D. C. Rivett, 27 November 1913.
50 MS. Dep. BAAS 147 [no pagination].
51 *The Times*, 27 August 1866.
52 National Library of Ireland, MS. 4792, Wicklow Papers: Journal of Caroline Howard [no pagination], 28 August 1857.
53 MS. Dep. BAAS 414 and *Irish Daily News*, 19 August 1878.
54 Mitchell Library, MS. M.P.31 D-TC, 4.1.31, fol. 510.
55 MS. Dep. BAAS 5, Miscellaneous Papers, 1831–1869, fol. 161, give attendance figures for annual meetings 1836–1850; a table (pp. xxxviii–xxxix) in the *Report of the Thirty-ninth Meeting of the British Association for the Advancement of Science* (London: British Association for the Advancement of Science, 1869) gives attendance figures (and receipts) for annual meetings 1831–1869: for some dates, the MSS. record and the 1869 table do not give the same figures (nor is the variation consistent between the two sources).
56 On diaries and female audiences at BAAS meetings, see R. Higgitt and C. W. J. Withers, 'Science and sociability: women as audience at the British Association for the Advancement of Science, 1831–1901', *Isis* 99 (2008), pp. 1–27.
57 Mitchell Library, MS. M.P. 31 D-TC, 4.1.31, fol. 512.
58 *Glasgow News*, 13 September 1876.
59 *Punch* 6 (1845), 25 June 1845.
60 *Punch* 3 (1842), pp. 6–7; 63 (1872), p. 77 respectively.
61 *Dublin University Magazine*, 4 September 1835.
62 *Freeman's Journal and Daily Commercial Advertiser*, 11 August 1835.
63 *Dublin Evening Post*, 25 August 1857.
64 'A word of welcome', *The Nation* XIV (1857), No. 52, p. 341.
65 *Irish Times*, 3, 10 September 1908.
66 Morrell and Thackray, *Gentlemen of Science*, p. 301.
67 Morrell and Thackray, *Gentlemen of Science*, p. 149.
68 National Library of Ireland, MS. 4792, 31 August 1857. The reporter to the *Dublin Evening Post* explained Livingstone's inaudibility thus: 'His voice had suffered severely from constant speaking under trees which had

no covering but the vault of heaven, and he regretted that he was not able to make himself better heard.' *Dublin Evening Post*, 1 September 1857.
69 *The Times*, 16 September 1853, pp. 5 f. (here referring to the 1853 Hull meeting).
70 Bodleian Library, MS. Eng.e.386, Diary probably by Agnes M. Hudson of her attendance at British Association meetings in Bristol and Sheffield and her journeys to France and Belgium, 1875–1879, fols 5v, 6–7, 11v–12v.
71 Bodleian Library, MS. Eng.e.386, fols 60v, 61v.
72 These questions have been more fully addressed in Higgitt and Withers, 'Science and sociability'.
73 S. Dierig, J. Lachmund and J. A. Mendelsohn, 'Introduction. Toward an urban history of science', *Osiris* 18 (2003), pp. 1–22.
74 D. Aubin, 'The fading star of the Paris Observatory: astronomers' urban culture of circulation and observation', *Osiris* 18 (2003), pp. 79–100, quote at p. 81.
75 D. Livingstone. 'Text, talk and testimony: geographical reflections on scientific habits: an afterword', *British Journal for the History of Science* 38 (2005), pp. 93–100, quote at p. 99.
76 On this point, see the chapters in M. Daunton (ed.), *The Organisation of Knowledge in Victorian Britain* (Oxford: Oxford University Press in association with the British Academy, 2005).

3

The science of geography in the British Association

Historians of geography have paid their subject's place in the British Association limited attention: historians of science and of the Association discuss geography hardly at all. From the vantage point of the centenary meeting in London in 1931, and in florid tones, the chemist-turned-geographer Hugh Robert Mill, one-time president of section E and, then, a vice-president of the Royal Geographical Society (RGS), offered the view that 'In the British Association the voice of Geography has been crying with a sound which is growing in clearness and certainty in these later days, although in the past it may sometimes have wavered or even sunk to inaudibility.'[1] In referring mainly to the period since 1885, Mill at least spoke from personal experience, having, as he put it, 'attended nearly all the meetings and held an official position in the section, either as member of committee, recorder, vice-president or president during most of the time'. For him, that half-century or so was a remarkable period in terms of section E's promotion of geographical research – in polar work, in survey and exploration, in study of Africa – and because of close links, in the 1880s and 1890s especially, between the BAAS and the RGS (even if he did also observe that 'During the earlier part of this period the annual gathering of Section E continued to be largely a summer meeting of the Royal Geographical Society').[2]

For Alice Garnett, recalling the role of section E in the 1920s, 'attendance at the September meetings of Section E came to carry greater importance for academic geographers than the January GA [Geographical Association] conferences'. Further, 'the encouragement afforded by these meetings for research, the reading of research papers, and discussions, with the added incentive of field days at the weekend, played an immeasurably important part in laying the foundations for our future independent development'.[3] Howarth's centennial review of section E adds little to Mill's bravura account. More recent authors either mention the Association only in passing or neglect it altogether, seeing geography's place in the RGS, in the universities or in schools as the only features of its civic presence and intellectual-institutional development in and from the nineteenth century.[4]

What was it that these geographer-scientists remembered? In examining the nature of geography as a science within the BAAS by considering the workings of section E, this chapter is concerned primarily with the cognitive content and epistemological reach of geography as a form of associational civic science, with what contemporaries understood by that term and with geography's different audiences and disputes over its scientific status. My approach, as throughout the book, is that of spatially sensitive social constructivism, a concern with the categories and meaning given to geography by contemporary actors, to 'the role of human beings, as social actors, in the making of scientific knowledge', and with the sites and spaces, social and material, in which geographical knowledge was made and received by different people.[5] It is also, in part, comparative in relation to what else is known about geography as a science in nineteenth- and early twentieth-century Britain.

Several questions are apposite. How was geography put to work within the BAAS? For Garnett, who saw in the activities of section E geography's professional foundations, and for Mill, whose memories of geography's 'voice' in the BAAS spanned nearly fifty years, what exactly was being recalled? How did section E work, on what and with whom? To answer these and other questions demands that we consider the cognitive content of what geography was held to be by its practitioners and its audiences and the epistemological practices, such as mapping, writing and exploring, through which what was taken to geography worked as a science and was debated by its several audiences. To these ends, the chapter is in four parts. The first considers geography's sectional history as section E, its place in the BAAS in terms of geography's funding relative to other sciences and the nature of its connections, personal, epistemological and political, with other sections. The second examines what sort of geography was being undertaken in and through the BAAS, and by whom, with reference to the papers given to section E meetings. The third section looks at the presidential addresses within section E, to disclose who amongst the more influential figures in the subject was promoting what sort of geography when. These addresses were declamatory acts and sites of speech. They were, often, calls for 'disciplinary' unity in the face of divergent tendencies and personal beliefs within geography between 'scientific' and 'exploratory' practices. As such, they provide insight into major differences within section E and within geography's public audiences over what the subject should include and how it was a science. The final part considers evidence for geography's differing reception for different audiences and in different places.

Science and geography: the nature and work of section E, 1851–1939

Geography's place as a science

Geography was initially placed within the Geological and Geographical Sub-committee (the title 'Section' was not used until 1833, initials not until 1835). It was initially combined with geology both as the result of the predominance of influential geologists within the formative years of the BAAS and because geography had no clear and separate subject focus of its own apart from geology.[6] Geography was identified as section E from 1851, designated then as 'Geography and ethnology', not 'Geography' alone. This remained the case until 1869, when ethnology briefly shifted to section D (biology) before becoming section H (anthropology). Sectional affiliation did not equate with epistemological equivalence nor promote scientific affability amongst fellow section members: as Chapters 5 and 6 make clear, the relationships between geographers and ethnologists between 1851 and 1871 (and even after) were often strained, as they were earlier with geologists, even when section E addresses covered material of mutual interest such as that by the orientalist and ethnologist John Crawfurd.[7] What lay behind the disputes in mid century and after between, on the one hand, geographers and the ethnologists-cum-anthropologists and, earlier and on the other hand, between geographers and other earth scientists is explored in more detail in the following chapters. Here it is sufficient to note that section E had its beginnings in 1851, and from that date we can more clearly see how the subject was put to work. The fact that for almost eighteen years thereafter it was combined as a section with those whose interests embraced the emergent human sciences, as for eighteen years beforehand it had been allied as a junior partner with the earth sciences, is indicative of the slow moves to subject identity in Victorian science and of the fact that clear boundaries between discrete subjects could not easily, if ever, be drawn.[8]

Funding the science of geography

Sections were managed by committees, whose members and officers were drawn from amongst leading practitioners of the science in question. The committees which at one time or another directed section E advanced geography in several ways: through grants in aid of research, via formal recommendations or resolutions which did not receive funding and in organising sessions of papers at the annual meeting. Formal recommendations without funding were directed, usually, at a branch of British government (for example, the Hydrographic Department of the Admiralty, or the Foreign Office), often in association with the RGS or

Table 3.1 Grants made in support of the work of the British Association for the Advancement of Science, by section, 1831–1931

Section	Total sum expended (to nearest £)	Total No. of grants	Section grants as % of total No. of grants	Mean size of grant (to nearest £)	Section grant funding as % of total BAAS funding awarded
A Mathematics and Physics	36,235	134	19.3	270	39.7
B Chemistry	4,070	72	10.3	56	4.4
C Geology	7,618	103	14.9	73	8.4
C, D, E and K[a]	735	3	0.4	245	0.8
D and K Zoology and Botany	1,484	15	2.2	99	1.6
D Zoology	14,000	87	12.5	160	15.3
E Geography	3,834	31	4.5	123	4.2
F Economics	1,422	25	3.6	56	1.6
G Engineering	4,233	24	3.5	176	4.6
H Anthropology	10,491	70	10.2	149	11.5
I Physiology	3,361	59	8.5	57	3.7
J Psychology	182	1	0.1	182	0.2
K Botany	2,879	44	6.3	65	3.2
L Education	709	25	3.6	28	0.8
M Agriculture	5	1	0.1	5	–
Total	91,218	694	100.0	Mean 57	100.0

Note:
a Research involving all of Geology, Zoology, Geography and Botany.

Source: Howarth (1931), appendix I, pp. 266–92.

the Royal Society, many of whose fellows were members of the BAAS and vice versa.

According to Howarth in his centennial record of the Association, thirty-one grants were made to geography in the century from 1831.[9] The totals for that subject and all those supported by the BAAS are shown in Table 3.1. What this evidence shows is that within the BAAS as a whole, and in relation to the nearly 1,000 grants awarded, funding for scientific research was dominated by the physical sciences of section A, mathematics and physics (134 grants), and by zoology (eighty-seven) and geology (103) (sections D and C, respectively). Several sciences had no more than two dozen or so grants in this period, even recognising that not all sections were of the same duration within the BAAS. Not including grants awarded for cross-sectional work, the average number of grants was fifty-two per section – that is, about two funded projects per year in the period 1831–1931, at a level, on average, of about £57 per grant. A note of caution is necessary in interpreting these figures: the relatively high proportion of funding for anthropology (section H) is misleading, for instance, since many grants were for cave exploration into hominid palaeontology in association with geologists. And interrogation of BAAS manuscript evidence reveals that additional funds were allocated although no new award or record of the additional money was noted by Howarth.

Geography's overall funding of £3,834 from thirty-one grants in the period 1831–1931 represents only a little over 4 per cent of the total number of grants and of the total funding allocated for science. Even so, geography fared better than seven other BAAS sections and sciences in terms both of the total sum expended upon grants and the mean size of grants awarded in the period. Relatively, however, the support offered by section E was negligible compared with the sums allocated by the RGS, which awarded eighteen separate grants over £500 between 1830 and 1914 – some, like Joseph Thomson's 1882–84 East Africa expedition, being awarded £3,595 – to a total of £30,681.[10] Some BAAS-funded geographical research was jointly undertaken with other sections – on the East African Lakes in the late 1920s, for example – and was so from recognition that different sciences brought different perspectives to regional problems, for example, or by virtue of the perceived benefits of collaborative applications when the mean level of funding within the BAAS was, to judge from the case of the RGS, lower than support from specialist subject bodies. Other sections at times incorporated studies of a geographical nature: geodetic investigations and African hydrography in mathematics and physics, and from sections D and K in 1841, for example, on 'The geographical distribution of animals and plants'. Personal connections sometimes sustained or influenced sectional relationships: Halford Mackinder's 'autobiographical fragments' show that his acquaintance with Thomas Elliot, secretary to section F (economics) and first permanent secretary to the Board of Trade, allowed the two sections to secure a joint grant.[11]

This 'overlap' in funding and cognitive content between section E and other sections and, thus, the blurred relationship between geography and other sciences illuminates the more general problem of the demarcation of the sciences one from another and, indeed, of 'science' from other forms of knowledge.[12] Assessment of the annual BAAS *Reports* and of manuscript records gets us a little closer to the detailed workings of the sciences, certainly to a clearer picture of funding levels than is afforded by Howarth. Between 1852 and 1936, fifty-six grants were made for geographical work, with a further twenty-two applications made and not funded and eight partially funded (Table 3.2). Some grants seem to have gone unused because, in the view of the intended recipient, the sum was too small even to allow a start on the work. The difference in totals between Howarth's summary list of thirty-one grants (Table 3.1) and the fuller listing of geography projects presented in Table 3.2 arises from the fact that Howarth did not include grants proposed after 1931, and because he listed only grants awarded. Some projects repeatedly sought funding through their organising committee before being supported, and one or two projects, although differently named, involved substantially the same personnel and objectives. This is true, for instance, of mapping and survey work in Central Moab, Palestine and the Near East. There, joint work between geographers in the BAAS with the Palestine Exploration Fund, the RGS and the Royal Engineering Corps in scriptural geography and in the study of biblical landscapes reflected shared institutional membership – for the later leading Ordnance Survey figure Charles W. Wilson amongst others – and mounting international cooperation in such questions.[13]

In the third quarter of the nineteenth century, section E funding was principally directed at terrestrial exploration, notably of New Guinea in the 1880s, but also toward Ernst Georg Ravenstein's work on the physical geography of Africa and in Greenland, Palestine and in the Karakoram mountains in northern Kashmir. Applicants and recipients were mainly leading men of science, military men or, increasingly from the 1890s, professional geographers. From about 1898, with the exception of some jointly funded topographic surveys, polar and oceanographic work was more to the fore, in the work of Sir John Murray (albeit that not all of his applications to section E were successful) and in bathymetrical charting in the Antarctic. The BAAS, like the RGS, was involved in funding later nineteenth-century Himalayan survey work, employing native men in the service of imperial delineation.[14] Topics in cartography and geography in education first appear in the early twentieth century, some of the work under the direction of Sir Charles Lucas, author of works on the historical geographies of the British Empire, and who was as we shall see a leading figure behind the BAAS visit to Australia in 1914 (Chapter 4). Much of the work of the section E Committee on the Teaching of Geography under Nunn's chairmanship was undertaken in collaboration with the Geographical Association. What is revealed

Table 3.2 Grants made to section E (geography), 1852–1936

Applicant	Year awarded	Topic	Sum awarded (to nearest £)
Murchison, Sir R. I.	1852	Large outline map of the world	15 [paid 1855]
Nicholson, Sir C.	1866	Palestine exploration	50
Murchison, Sir R. I.	1867	Greenland exploration	100
Murchison, Sir R. I.	1871	Exploration of the Country of Moab	100
Wilson, Major C.	1874	Palestine Exploration Fund	100
Council of the BAAS	1881	Exploration of the mountain district of eastern equatorial rain forest [New Guinea]	100
Godwin-Austen, Lt Col.	1883	Exploration of New Guinea	100 [not paid]
Lefroy, Gen. Sir H.	1884	Exploration of New Guinea	200
Lefroy, Gen. Sir H.	1884	Exploration of Mt Roraima [British Guiana]	100
Walker, Gen. J. T.	1885	New Guinea exploration	150
Walker, Gen. J. T.	1885	Depth of permafrost in polar regions	5 [not paid]
Walker, Gen. J. T.	1886	Bathy-hypsographical map	
Walker, Gen. J. T.	1886	Depth of permanently frozen soil	5 [not paid]
Walker, Gen. J. T.	1887	Depth of permanently frozen soil	5
Walker, Gen. J. T.	1888	Geography and geology of the Atlas Mts	100
Garson, Dr	1890	Nomad tribes of Asia Minor and northern Persia	30 [25 paid]
Ravenstein, E. G.	1891	Climatology and hydrography of tropical Africa	75 [50 paid]
Godwin-Austen, Col.	1892	Exploration of Karakorum Mts	50
Wilson, Sir C. W.	1892	Scottish place names	10
Ravenstein, E. G.	1892	Climatology and hydrography of tropical Africa	50
Ravenstein, E. G.	1893	Climatology and hydrography of tropical Africa	10 [not paid]

The science of geography 73

Table 3.2 (continued)

Applicant	Year awarded	Topic	Sum awarded (to nearest £)
Markham, C. R.	1893	Observations in South Georgia	50
Seebohm, H.	1893	Exploration in Arabia	30
Ravenstein, E. G.	1894	Climatology of tropical Africa	5
Seebohm, H.	1894	Exploration of Hadramout [Arabia]	50
Ravenstein E. G.	1895	Climatology of tropical Africa	10
Ravenstein, E. G.	1896	Climatology of tropical Africa	20
Ravenstein, E. G.	1897	Climatology of tropical Africa	10
Keltie, J. S.	1898	Exploration of Socotra	35
Murray, Sir J.	1899	Physical and chemical constants of sea water	100
Keltie, J. S.	1900	Terrestrial surface waves	5
Mill, H. R.	1900	Changes of land level in the Phlegraean Fields	50
Keltie, J. S.	1901	Terrestrial surface waves	15
Keltie, J. S.	1902	Tidal bore, sea waves, and beaches	25
Holdich, Sir T. H.	1902	Scottish National Antarctic Expedition	50
Murray, Sir J.	1904	Investigations in the Indian Ocean	150
Murray, Sir J.	1905	Rainfall and lake and river discharge	10 [not paid]
Hogarth, D. G.	1906	Oscillations of the land level in the Mediterranean basin	50]
Murray, Sir J.	1906	Rainfall and lake and river discharge	10
Murray, Sir J.	1907	Investigations in the Indian Ocean	50
Murray, Sir J.	1907	Rainfall and lake and river discharge	5 [not paid]
Chisholm, G. G.	1907	Exploration in Spitsbergen	30
Murray, Sir J.	1908	Investigations in the Indian Ocean	35

Table 3.2 (continued)

Applicant	Year awarded	Topic	Sum awarded (to nearest £)
Murray, Sir J.	1908	Rainfall and lake and river discharge	10 [not paid]
Chisholm G. G.	1910	Map of Prince Charles Foreland	30
Herbertson, A. J.	1910	Equal area maps	20 [not paid]
Herbertson, A. J.	1911	Equal area maps	20 [not paid]
Close, Col. C.	1911	Calculation of areas on the spheroid	25 [not paid]
Myres, J. L.	1913	Maps for school and university use	40
Dickson, H. N.	1913	Tidal currents in Moray and adjacent firths	40
Lucas, Sir C. P.	1914	Conditions determining selection of sites and names for towns	20
Chisholm, G. G.	1914	Survey of Stor Fjord, Spitsbergen	50
David, T. W. E.	1914	Antarctic bathymetrical chart	100
Lucas, Sir C. P.	1915	Conditions determining selection of sites and names for towns	15
Nunn, T. P.	1922	Syllabus for the teaching of geography	10
Beckit, H. O.	1923	Provisional population map of the British Isles	5 [not paid]
Nunn, T. P.	1924	Syllabus for the teaching of geography	5 [not paid]
Beckit, H. O.	1924	Provisional population map of the British Isles	30 [not paid]
Nunn, T. P.	1926	Syllabus for the teaching of geography	5
Beckit, H. O.	1926	Provisional population map of the British Isles	25 [not paid]
Lucas, Sir C.	1926	Geographical knowledge of tropical Africa	5 [not paid]
Nunn, T. P.	1927	Syllabus for the teaching of geography	5 [not paid

The science of geography 75

Table 3.2 (continued)

Applicant	Year awarded	Topic	Sum awarded (to nearest £)
Beckit, H. O.	1927	Provisional population map of the British Isles	30 [25 paid]
Lucas, Sir C.	1927	Geographical knowledge of tropical Africa	10 [not paid]
Nathan, Sir M.[a]	1927	Biology, geology and geography of the Great Barrier Reef	400
Fawcett, C. B.[b]	1927	Soil resources of the empire	?
Beckit, H. O.	1928	Provisional population map of the British Isles	50
McFarlane, J.	1928	Geographical knowledge of tropical Africa	25 [not paid]
Nunn, T. P.	1928	Syllabus for the teaching of geography	5 [not paid]
Beckit, H. O.	1929	Provisional population map of the British Isles	75 [not paid]
Roxby, P. M.	1929	Geographical knowledge of tropical Africa	50 [26.15 paid]
Nunn, T. P.	1929	Syllabus for the teaching of geography	5 [3.4 paid]
Gardiner, J. H.[c]	1930	Investigation of Lakes Baringo and Rudolph, northern Kenya	200
Winterbotham, H.	1930	Population map of Great Britain	25 [not paid]
Roxby, P. M.	1930	Geographical knowledge of tropical Africa	10 [2.65 paid]
Nunn, T. P.	1930	Syllabus for the teaching of geography	4 [not paid]
Close, Sir C.	1931	Population and food supply	30
Roxby, P. M.	1931	Geographical knowledge of tropical Africa	25 [20.53 paid]
Fawcett, C. B.	1931	Geography in Dominion universities	?
Roxby, P. M.	1932	Geographical knowledge of tropical Africa	5
Russell, Sir J.[d]	1932	1:M scale maps of woodland in Britain	25
Kitson, Sir A.[e]	1933	Topographical and geological survey of the	35

Table 3.2 (continued)

Applicant	Year awarded	Topic	Sum awarded (to nearest £)
Roxby, P. M.	1933	Lake Rudolph area, East Africa Geographical knowledge of tropical Africa	5
Russell, Sir J.	1933	1:M scale maps of woodland in Britain	25
Debenham, F.	1935	Land forms of the North East Land [Spitsbergen]	25
Rudmose-Brown, R.	1936	Insolation and population	25

Notes:
a On behalf of sections D (zoology), E (geography) and K (botany).
b On behalf of sections E (geography) and M (agriculture).
c On behalf of sections C (geology), D (zoology), E (geography) and H (anthropology).
d On behalf of sections E (geography) and K (botany).
e On behalf of sections C (geology) and E (geography).

Source: BAAS, *Reports of Meetings* and BAAS Archives, MS. Dep. BAAS *passim*.

is a detailed campaign to secure geography's recognition as a science in schools at a time when concerns were being expressed over study of science as a basis to citizenship[15] (questions explored more fully in Chapter 7).

The pattern of funded geographical work supported by section E is mirrored in the record of recommendations and resolutions from 1852. Although section E recommendations did not receive grants of money, they often led to applications for funding: they were more statements of intent, written representations of the significance of certain geographical topics. Between 1852 and 1932, seventy-nine recommendations or resolutions were advanced with, before c. 1900, an emphasis on polar exploration, terrestrial and oceanographic exploration and several stressing the overall utility to science of standard forms of and uniform scales in mapping (see also Chapter 5). Cross-sectional collaboration was a noticeable feature of later recommendations. Of thirty-nine recommendations proposed between 1908 and 1936, fifteen were in association with other sections, notably with botany, zoology and mathematics on, respectively, ecological and geodetic survey. In terms of the working relationship with the RGS, the two bodies were often co-proponents of geographical resolutions – in Palestine and in the Himalayas as we have seen, and over exploration of the River Niger in 1856, or in 1873–74 where the BAAS and the RGS came together with other learned bodies and civic commercial interests to promote Arctic exploration.[16] But only a handful of projects funded by the BAAS were published in the outlets of the RGS, its *Journal* or *Proceedings*, or, later, the *Geographical Journal*. The fact that one or two – Wilson's work in Palestine and Ravenstein's on hydrographic mapping in association with Walker – were published in RGS journals before they were supported by the BAAS (see Table 3.2) suggests that, in some cases, the Association was promoting already established geographical research rather than always funding new initiatives.[17] Generally, work in what we may think of as 'primary topographic survey' – large-scale mapping or expeditionary work in, for example, Africa, Tibet or Arabia – was to the fore until about 1900. Thereafter, although such work was still successfully funded, funding was more often directed at smaller-scale thematic projects.

From exploration to 'scientific geography': changes in nature and status

Geography's making: the programmes of section E

For the great part of the nineteenth century and in part into the early twentieth century, section E programmes were popular amongst the public attending BAAS annual meetings, principally because geography did not require detailed prior knowledge to be understood or enjoyed.[18] As we shall see in more detail in Chapter 6, one of the arguments in the debates over the separation of ethnology from geography was that section E meetings were so dominated by popular geography talks that ethnology had 'no chance' to develop a reputation through the BAAS without its own sectional identity.[19] Africa and the North Pole – 'the two main dishes' as *The Guardian*'s reporter put it in 1881 – could fail to attract their 'usual crowd of listeners', however, if, as happened that year, the explorers themselves failed to give the talk.[20]

In order to know better what it is that people were commenting upon, we can look at what was being delivered, and how, in the programme of papers presented at section E meetings. Before 1851, most 'geography' papers were in physical geography or in related topics with the occasional exception, such as the paper in 1846 by the Irish-born Africanist William Cooley on the desirability of a 'physico-geographical survey of the British Islands' where he argued for the importance of

Table 3.3 Content and number of geography papers given to section E meetings, by topic and decade, 1851–1933

Content	Decade							
	1851–1860	1861–1870	1871–1880	1881–1890	1891–1900	1901–1910	1911–1920	1921–1933
Acclimatisation	–	–	–	3	10	2	–	–
Cartography	9	2	2	10	6	5	10	7
Climatology	5	6	2	6	5	5	3	10
Commercial geography	5	5	7	16	11	17	2	25
Engineering	9	3	1	7	–	–	1	4
Ethnology	–	2	2	1	4	2	4	8
Exploration	87	84	123	124	91	36	17	1
General[a]	–	–	4	10	13	11	6	8
Geographical education	–	–	2	7	8	6	5	7
Geology (including geodesy)	1	–	2	1	2	2	1	1
Geomorphology	–	–	–	–	10	30	16	21
Glaciology	–	2	1	–	4	5	1	2
Historical geography	1	4	3	8	7	6	5	22
Hydrology	16	–	10	2	4	2	2	2
Limnology	1	–	2	4	2	3	1	–
Meteorology	1	–	–	–	6	1	2	1
Mountaineering	3	1	1	1	2	–	–	–
Navigation	1	6	8	2	–	2	–	3
Oceanography	8	4	11	9	26	1	4	3
Physical geography (general)	2	17	15	12	8	8	4	5
Plant geography	2	1	2	–	–	8	–	–
Political geography	–	–	1	2	5	3	9	5
Regional geography	–	–	–	–	1	6	12	20
Seismology and volcanology	5	4	1	2	7	5	–	1
Social geography	–	–	1	1	1	2	3	28
Survey	–	–	6	14	13	9	9	20
Topography	3	1	4	–	1	1	3	2
Travel	–	–	1	–	1	7	–	7
Urban geography	–	–	–	–	–	1	2	7
(Other)[b]	–	–	3	1	2	2	3	–
Total	159	142	215	243	250	189	125	220

Notes:
a This category includes all papers of a very general nature, neither regionally nor thematically specific.
b This category includes papers on place names, geographical terminology, one on medical geography and two on population distribution.

Source: *Reports of the Meetings*; MS. Dep. BAAS, *passim*.

integrating local climate data in improving agricultural science.[21] Table 3.3 shows the content of the geography papers delivered in the period 1851–1933. Here, too, questions of demarcation arise and some degree of caution is necessary in interpretation: the sub-categories, drawn up after reading the papers and using then contemporary classifications, are not always mutually discrete (where 'exploration' as expeditionary science also involved, say, topographical surveying or navigation). Much early historical geography – if so described by speakers – might as well have been described as 'scriptural geography' given its attention to the historical geography of the Holy Land in the collaborative surveying with the RGS and the Palestine Exploration Fund. Some papers within 'commercial geography' – the economic benefits of international trade in an imperial context – were strongly political in content. 'Engineering' embraced elements of railway geography and could as well have been included as transport geography. The increased frequency of the term 'geomorphology' reflects contemporaries' more specific use of that term after about 1900 in assessment of the Earth's physical features – coastal processes, karst, dune and wave formation in sand and polar environments and so on – than the category 'Physical geography (general)'.

Several features are noteworthy in analysis of Table 3.3. Papers on exploration dominated BAAS section E meetings until about 1900 but hardly featured by the 1920s. Where grant funding for exploratory geography continued, albeit to a lesser extent, after 1900, the presentation of such topics in BAAS meetings seems to have taken longer to decline and to have begun earlier. Consider the drop in exploration papers from a high point of 124 papers in the years 1881–90, for example, to ninety-one by 1891–1900 to seventeen by 1911–20: Table 3.3. Where in the twenty years from 1881 to 1900 papers on exploration represented nearly 44 per cent of all papers delivered at BAAS meetings, they made up less than 17 per cent in the following twenty years, and less than 1 per cent by the 1920s. In general terms, this evidence on papers presented corroborates the evidence on grants (Table 3.2) which suggests a decline in the number of awards given to exploration from about this date and a relative rise in more thematic topics. Yet this demarcation of one sort of geography from another is not clear-cut. In place of a focus on exploration, a more diverse geography was evident. Commercial geography was more to the fore, its emphasis trade and the economics of natural resources often in regard to the British Empire. Its rise in the period 1921–33 reflected engagement with the economic development of Britain's colonies (which, as we shall see in Chapter 4, was discussed for Canada at the 1884, 1897, 1909 and 1924 BAAS meetings, for Australia in 1914 and for South Africa in 1905 and 1929). The growth in regional geography can in part be explained by papers on local geographies produced for the Association's later handbooks, effectively texts of regional survey (see Chapter 7). More clearly defined work in social geography, urban geography and in physical geography, in geomorphology and in survey (whose increase in the 1920s reflected interests in aerial survey as a geographical technique and in the larger-scale institutionalisation of national mapping programmes under governmental direction) are also distinctive elements.

Taking the evidence of the grants and the papers together, and allowing for some slippage between the categories employed by contemporaries, there is a clear sense of an emergent disciplinary specialisation within the science that was geography in the BAAS, but at different times and perhaps more clearly for the spoken papers than the funded projects. Because the geography promoted within the BAAS was never a single or a simple thing, we must be cautious about seeing in the work of section E and in its outcomes either a clear and unequivocal 'transition' towards geography's 'modernity' or clean lines of epistemic demarcation between geography and (say) anthropology, geology, economics (in terms of work on trade and commercial exchange), even biology in terms of interest in the distribution of plant and animal communities. The fact that 'exploration' and 'mapping' continued to be supported in one context but less so in another – and in the meetings later than in the grants awarded – highlights the importance of considering the different ways in which the BAAS worked.

Sites and acts of 'disciplinary' declaration: the presidential addresses of section E

Consideration of the presidents' addresses complicates the picture. Table 3.4 lists the subject matter and the speakers of the section E presidential addresses between 1858 and 1939.[22] On occasion, the geography section was without 'professional' geographers to give its sectional address: given his significant role on the voyage of the *Challenger* (1872–76), for example, the oceanographer Charles Wyville Thomson was asked by BAAS Council to preside over the 1878 meeting – an opportunity he used to criticise the African explorer Henry Morton Stanley for the reported violence of his latest explorations (Figure 3.1). In York in 1881, the botanist Sir Joseph Hooker spoke (in the absence of a 'professed geographer' as he put it) upon the geographical distribution of plants and animals and the history of botanical geography.[23] For those leading men of geography and of science who did give the section E presidential

The science of geography 81

Table 3.4 Section E presidential addresses, 1858–1939: speaker and subject

Year	Location	Speaker	Principal subject matter
1858	Leeds	Sir R. I. Murchison	Relationships between geography and ethnology
1860	Oxford	Murchison	African exploration; definition of geography
1861	Manchester	Sir John Crawfurd	Ethnology and physical geography
1863	Newcastle upon Tyne	Murchison	Geography and exploration; role of Section E
1864	Bath	Murchison	Geography and exploration; role of section E
1865	Birmingham	Sir Charles Nicholson	Geography and ethnology; exploration
1867	Dundee	Sir Samuel Baker	Geography, exploration and commerce
1868	Norwich	Capt. Richards, RN	Geography and exploration
1869	Exeter	Sir Bartle Frere	The progress of geographical discovery
1870	Liverpool	Murchison	Geography and the role of section E
1871	Edinburgh	Sir Henry Yule	Physical geography of Far East
1872	Brighton	Francis Galton	Nature of section E; OS mapping
1873	Bradford	Sir Rutherford Alcock	Scope and application of geography
1874	Belfast	Maj. Charles Wilson	Physical geography and military operations
1875	Bristol	Lt Gen. R. Strachey	Geography, exploration and scientific specialisation
1876	Glasgow	F. J. Evans	African exploration; *Challenger* expedition
1877	Plymouth	Sir Erasmus Ommanney	Advances in geographical science since 1841 [last Plymouth meeting]
1878	Dublin	Sir C. Wyville Thomson	*Challenger* expedition
1879	Sheffield	Clements Markham	'The objects and aims of geographers'
1880	Swansea	Lt Gen. Sir J. H. Lefroy	Exploration in North America
1881	York	Sir Joseph Hooker	Geographical distribution of animals and plants

Table 3.4 (continued)

Year	Location	Speaker	Principal subject matter
1882	Southampton	Sir Richard Temple	Central plateau of Asia
1883	Southport	Lt Col. H. Godwin-Austen	Physical geography of the Himalayas
1884	Montreal	Lefroy	Recent progress in geographical science
1885	Aberdeen	Gen. J. T. Walker	Scientific geography and mapping
1886	Birmingham	Maj. Gen. F. J. Goldsmid	Popularising geography through schools
1887	Manchester	Col. Sir C. Warren	Geography education in schools
1888	Bath	Col. Sir C. W. Wilson	Geography and trade; exploration
1889	Newcastle upon Tyne	Sir Francis de Winton	'Science of applied geography, commerce'
1890	Leeds	Lt Col. Sir Lambert Playfair	The Mediterranean
1891	Cardiff	E. G. Ravenstein	'The field of geography'
1892	Edinburgh	James Geikie	Geography of coast lines
1893	Nottingham	Henry Seebohm	Nature of geography; polar exploration
1894	Oxford	Capt. W. Wharton	Hydrography
1895	Ipswich	H. J. Mackinder	Geography, exploration and university education
1896	Liverpool	Maj. L. Darwin	Polar exploration; geography and trade
1897	Toronto	J. Scott Keltie	Exploration; geographical education
1898	Bristol	Col. G. Earl Church	South America
1899	Dover	Sir John Murray	*Challenger* expedition; oceanography
1900	Bradford	Sir G. S. Robertson	'Political geography and the empire'
1901	Glasgow	H. R. Mill	On research in geographical science'
1902	Belfast	Col. Sir T. H. Holdich	Geography, geodetic survey
1903	Southport	Capt. E. Creak	'Terrestrial magnetism and its relation to geography'

Table 3.4 (continued)

Year	Location	Speaker	Principal subject matter
1904	Cambridge	Douglas W. Freshfield	'On mountains and mankind'
1905	Cape Town	W. J. L. Wharton	'The field of geography and some of its problems'
1906	York	G. T. Goldie	'Twenty-five years' geographical progress'
1907	Leicester	G. G. Chisholm	'Geography and commerce'
1908	Dublin	Maj. E. H. Mills	The present and future work of the geographer'
1909	Winnipeg	D. Johnston	'The survey and mapping of new areas'
1910	Sheffield	A. J. Herbertson	'Geography and some of its present needs'
1911	Portsmouth	Col. C. F. Close	'The position of geography'
1912	Dundee	Col. Sir C. M. Watson	Geography, exploration, role of R. I. Murchison
1913	Birmingham	Prof. H. N. Dickson	Geography, exploration, regional description
1914	Australia	Sir Charles P. Lucas	'Man as a geographical agency'
1915	Manchester	H. G. Lyons	'The importance of geographical research'
1916	Newcastle	E. A. Reeves	'The mapping of the earth, past, present and future'
1919	Bournemouth	Prof. L. W. Lyde	'The international rivers of Europe'
1920	Cardiff	J. McFarlane	Political geography; boundaries after World War I
1921	Edinburgh	D. G. Hogarth	'Applied geography'
1922	Hull	Marion I. Newbigin	'Human geography: first principles and some applications'
1923	Liverpool	Dr V. Cornish	'The geographical position of the British Empire'
1924	Toronto	Prof. J. W. Gregory	'Inter-racial problems and white colonisation of the tropics'
1925	Southampton	A. W. Hinks	'The science and art of map-making'
1926	Oxford	W. Ormsby-Gore, MP	'The economic development of tropical Africa and its effect on the native population'

Table 3.4 (continued)

Year	Location	Speaker	Principal subject matter
1927	Leeds	R. N. Rudmose Brown	'Some problems of polar geography'
1928	Glasgow	J. N. L. Myres	'Ancient geography in modern education'
1929	South Africa	Brig. E. M. Jack	'National surveys'
1930	Bristol	Percy M. Roxby	'The scope and aims of human geography'
1931	London	Sir H. J. Mackinder	'The human habitat'
1932	York	Prof. H. J. Fleure	'The geographical study of society and world problems'
1933	Leicester	Lord Meston	'Geography as mental equipment'
1934	Aberdeen	Prof. A. G. Ogilvie	'Co-operative research in geography: with an African example'
1935	Norwich	Prof. Frank Debenham	'Some aspects of the polar regions'
1936	Blackpool	Brig. H. S. L. Winterbotham	'Mapping of the colonial empire'
1937	Nottingham	Prof. C. B. Fawcett	'The changing distribution of population'
1938	Cambridge	Griffith Taylor	'Correlations and culture'
1939	Dundee	A. Stevens	On natural geographical regions

Notes: Where a year is missing, no section E presidential address was given. Many section E presidential addresses were untitled: where that is so, the subject matter has been identified by close reading of the address in question.

Source: *Report of the Meeting* [of the BAAS].

addresses, the principal subjects covered were geography and exploration (with different emphases at one time or another upon Africa, the Americas, Asia or the polar regions); geography and ethnology (in terms both of the subject matter of human cultures and their environmental bases, and the strained intellectual and social relations between proponents of the two subjects in the BAAS: see Chapter 5); mapping as a form of imperial and national survey; geography and education; and, least often, the remit of section E itself as a forum for promoting geography.

Several addresses were notable in terms of the strident tones

Figure 3.1 Sir Joseph Lister giving the presidential address at the BAAS meeting, Liverpool, 1896

taken by the speaker in emphasising his topic. At the 1864 Bath meeting, for instance, where a crowd of thousands gathered to hear David Livingstone speak on his return from the 1858–63 Zambesi Expedition, Murchison used his presidential address to counter views that geography was 'exhausted' since so few places remained to be known. Exploration was key to geography, he emphasised. Moreover, reports upon exploration were key to the popularity of geography: understood thus, geography is a 'noble and unlimited science'.[24] Eight years later, in Brighton, where Livingstone again attracted large crowds to section E, Francis Galton (twice a president of the section) argued forcefully against this view, noting that 'The career of the explorer, though still brilliant, is inevitably coming to an end.' If the section was 'to secure the attention of representatives of all branches of science', it was essential that geography turn to 'principles and relations' rather than 'primary facts'.[25] This point about geography's credentials as a science (which ought to be apparent in its methods) and its connections with other sciences was emphasised by Strachey in his 1875 presidential address. It is a key feature of the addresses in the later 1870s. In stressing that a 'more scientific geography' should incorporate 'the doctrine of evolution' in the study of human–environment relationships, Strachey sought both to situate geography in relation to wider debates and to improve geography: 'the field of topographical exploration is already greatly limited . . . The necessary consequence is an increased tendency to give geographical investigations a more strictly scientific direction'.[26]

Even, then, as section E sanctioned and funded exploration (*cf.* Table 3.2), and as BAAS geographical audiences warmed to such work, many of its leading officers were publicly proclaiming the end of that tradition of geographical enquiry. For Strachey, endorsing in his own 1875 speech what Galton had forcefully expounded three years before, this need not mean that the BAAS should cease funding exploration but that explorations should be more scientific. In turn, 'scientific geography' should be part of better science: the 'general progress of science' must involve 'the study of geography in a more scientific spirit'.[27] Private correspondence reveals that Strachey's public pronouncements were listened to, at least by some other geographers. Writing to his mentor H. W. Bates, the mapmaker and Africanist Alexander Keith Johnston noted 'As General Strachey pointed out in his address at Bristol in 1875, 'the geography of the future must take a new channel. It can only take two directions – that of examining more minutely the topography of the lands and seas, the general natures of which are already known; – or it may turn to the solution of the problems of scientific geography'. Johnston opined that 'The former direction promises no immediate sources of interest for the general mass of people. . . to give attractiveness there must be something unknown; . . . but the latter path leads into an inexhaustible field of research and discovery of a higher order than that of the outline of new lands'.[28]

What did the phrase 'scientific geography' mean? Contemporaries' claims centred upon the role of detailed examination, systematic explanation and, increasingly, thematic focus coupled with rigour in method, rather more than they did upon more detailed topographic survey, advances in exploration or in regional description. In his address in Sheffield in 1879, the leading geographer and India Office civil servant Clements Markham endorsed Strachey's message and Johnston's sentiments by distinguishing three classes of geographical worker: the geodesist, the topographical surveyor and the trained explorer. Although as Markham put it there was work to be done, in 'raising the status' of the first 'to the heroic status of their exploring brethren', what mattered in producing a 'full-blooded geography' was going beyond survey as an end in itself.²⁹ In planning the 1879 meeting, Markham worked to attract a variety of papers in order to secure just such a geography – and incurred the wrath of a leading hydrographer who thought his work diminished by Markham's vision.³⁰ A quarter of a century later and in addressing the 1906 York meeting, George Goldie looked back on twenty-five years of 'geographical progress' with reference to the growth of scientific method in geography and what he termed the 'democratisation of geography' in that period. Twenty-five years from now, he argued, 'our countrymen will be found to occupy the same position in the front rank of scientific geographers that their forefathers held in pioneer exploration'.³¹

The evidence of Table 3.3 would seem to bear him out. Two other pieces of evidence do not. The first is that moment of 'geographical heresy' in 1911 when, at a time of suggested sectional restructuring within the BAAS, as part of still-evident tensions between scientific and other forms of geography from within the discipline's leading practitioners and in relation to geography's still uncertain place as a science of citizenship and civic education, the then section E president, Charles Close, denounced the subject's scientific credentials. I return to this in more detail in Chapter 7. The other relates to the public's response to the different sorts of geography and the different sorts of speakers within section E. Given as they were by men with geographical interests and credentials – 'professed geographers' to use Hooker's term but not, in the modern sense of the term 'professional' geographers – section E presidential addresses made important spoken statements, using varying specific content, about the nature and focus of geography's scientific basis and its connections with other scientific work.

The tension between geography's scientific credentials and its identification with exploration is particularly apparent as a recurrent theme. It is tempting to see Murchison's advocacy of exploration, in Oxford in 1860 and in Bath in 1864 for example, and others' later denial of this focus as evidence of a shift towards 'modernity' in geography. But precisely because this tension is recurrent, we should think less of single 'moments' of modern disciplinary emergence and, at least as far as the presidential addresses are concerned, rather more of relative emphases which reflect personal views and interests more than they do disciplinary shifts. Strachey's address in 1875 was a powerful instance in the framing of a new direction for geography. It had later parallels in the addresses of Mill in 1901, and of Mills in 1908. But in Dundee in 1912 – one year after a presidential denouncement of geography's scientific status – the then president, Colonel Sir C. M. Watson, abstained from the then prevalent view 'that exploration in the traditional sense is at an end', and appealed to the memory of Murchison in doing so.

If the performance and content of such disciplinary statements varied, so did reception of them. Engagement with the message could even be affected by the simple facts of timing. Reporting in 1887 upon what he saw as the 'universal evil' of the lack of planning in conducting sectional proceedings, H. R. Mill made the problem clear:

> The long sittings, from eleven to three, and the impossibility of judging at what time a given paper will be taken, are productive of much dissatisfaction and annoyance. Only people of very great importance can command an audience in the section rooms between one and two o'clock; and there was hardly a day in any section on which the papers coming late on the list could be read properly, or discussed at any length. This was felt to be unjust, particularly by foreign members, who regarded the discussion of papers as the most important purpose of the Association. Some simple arrangements would greatly improve matters . . . . it would also prevent the undignified devouring of sandwiches by the chairman while papers are being read. . . . The arrangements for garden parties, conversaziones, and excursions were, as they have been for many years, much more perfect than those for the scientific meetings.[32]

Two years later, reporting upon the Newcastle meeting, Mill struck a different note. Although the 1889 meeting of section E was designed to help promote commercial themes and the role of provincial geographical societies in this and in similar towns,[33] neither the papers nor the subsequent discussion met those objectives.

> The address of Sir Francis de Winton as President was intended to strike a key-note of Commercial Geography, which it was hoped would give tone to the following papers and discussions. The result was disappointing. Except in one or two instances the papers which made any pretensions to the title of Commercial Geography contained very little Geography, and the discussion on several occasions became purely political, the addresses of some speakers being thoroughly inappropriate to a scientific meeting.

In other respects, however, the section's business was successful and warmly received:

> Explorers always attract the public, and this year they were well represented. Dr. Nansen as he recounted his adventures on the Greenland ice must have felt that his heroic journey had been fully appreciated. The simple manly style in which he told the now familiar story, and his quaint and unexpected gleams of humour, commanded the closest attention of a crowded and enthusiastic audience.[34]

Mill's observations spoke to an issue key to section E and to the claims of those who would dismiss geography's exploratory credentials and/or advance it as a science. Popular lectures on exploration – just the topic that leading men of science saw as detracting from geography's scientific status – were well attended, especially if given by those the public took to be leading figures, such as David Livingstone or Fridtjof Nansen. More specialist papers and presidential addresses which contributed to scientific knowledge – or so it was hoped – provided insight into geography's disciplinary emergence in Britain well before the establishment of formal teaching programmes for the subject. But the fact remains that many in geography's audiences were engaged less with geography's epistemic and institutional status as a science and much more with the spectacle of its exploratory cultures and the personal status of its expository speakers.

Siting geography's audiences

This sense that explorers in person, humour and a 'manly style' attracted audiences to geography, as did the visual cultures associated with the practices of telling and displaying expeditionary endeavour, is apparent from a report of the 1894 Oxford meeting:

> The attendance was from first to last exceptionally good, and the audiences held together with quite remarkable constancy. That, however, was evidently due to the popular rather than to the strictly scientific character of the papers and discussions. It must always be remembered that in the British Association two audiences have to be catered for – the scientists and the amateurs – and that the financial success of the meeting depends much more on the latter than on the former. Complaints are often made – they have been made this year – that the treatment of the subjects in some sections is so technical that it repels the lay members, who form the bulk of the membership of the association. The objection may or may not be deserving of notice; but it certainly does not apply to the Geographical Section. That section is the happy hunting-ground of the unattached and amateur Associate. Thanks to the profuse and promiscuous use of the magic lantern, it has become the attractive show-room of the Association. But geography as a science, or the scientific aspect of geography, does not gain much by this ephemeral popularity. The audience is panting for sensations; the ubiquitous and irrepressible globe-trotter is the ideal of the hour, and the sensation is all the greater if the globe-trotter happens to be a woman. The paper which attracts a crowded audience may be a tedious narrative of a holiday spent in Armenia, or in Mexico, or in the desert of Libya, or in Montenegro, or in Arabia; but

Figure 3.2 Portrait of Sir John Franklin

all its sins are forgiven if it is illustrated by what the official programme calls 'optical projections', which means, in common parlance, 'lantern slides'.[35]

In one interpretation, here is clear endorsement of the views of Galton and Strachey decades before. In another, popular engagement with geography was really with geography's heroic figures, an engagement all the stronger of course if the tale was told by the explorer or if heroic death in doing geography should add poignancy to the figure's later reception. Public interest in the fate of the polar explorer Sir John Franklin, for example, was especially apparent at the 1860 Oxford meeting in a public lecture on the lost polar expedition, displays of maps showing the limits of polar exploration and a public reading in the university's Sheldonian Theatre of a poem, specially composed by a Fellow of Trinity College, 'upon the life and character of Sir John Franklin with special reference to time, place, and discovery of his death'[36] (Figure 3.2).

In thinking about geography's different audiences, however, there are interpretative dangers with respect to the different types of evidence. Reports on science did not always see it as necessary to comment upon the rhetorical style of the scientist, and only occasionally did so with respect to the reaction of the audience. Reports on BAAS section E proceedings were regularly carried in geography journals, notably in the *Geographical Journal* (the outlet of the Royal Geographical Society) and in the *Scottish Geographical Magazine* (the outlet of the Royal Scottish Geographical Society). These journals often also carried the published versions of section E papers, as did the journals of other subjects and the transactions of numerous regional societies.[37] Clements Markham used his short-lived and popular *Geographical Magazine* (1874–78) to carry notices of the BAAS geography meetings.

Most published reports on BAAS meetings were short, summarising the president's speech and the content of sectional papers delivered. On topics that were the subject of repeated concern – exploration, or ethnology, or of more immediate local interest, given connections between the town in question and the content of the lecture programme – societies' officers would sometimes offer additional comment concerning the tone of the address, the reaction to presented papers and, occasionally, glimpses of the social dynamic underlying section meetings. Reporting upon the 1887 Manchester meeting, H. R. Mill offered no view upon Sir Charles Warren's presidential address other than to complain that the concurrent timing of most sections' presidential addresses meant that one could not learn of other sections' work. But he praised Sir Francis de Winton's public evening lecture on African exploration as 'a model of what a popular lecture should be. He handled his familiar theme with a systematic clearness and quiet enthusiasm which secured the sympathy and unfaltering interest of his large audience.' Mill also noted how the meeting 'was of more than usual interest to geographers, for, besides a number of valuable papers read to section E, there were many bearing directly upon the subject submitted to other sections'. Noting too that 'Very important work was got through on Monday' – which included discussions on the scope and teaching of geography and on the work of the Ordnance Survey – Mill reports upon scenes of 'extremely animated' debate and hurried papers because of lax timetabling. In closing, he criticised the poor quality of the BAAS handbook: 'a sorry contrast both in bulk and quality to the Birmingham hand-book of last year'.[38]

Newspaper reports in contrast are plentiful for many BAAS section E meetings. With other evidence, they disclose a public involvement not apparent elsewhere. Recall Lady Caroline Howard and her companions' response to Livingstone at the 1857 Dublin meeting. Livingstone attracted a huge crowd but spoke in a whisper: Lady Howard's companions saw but did not hear geography's figurehead. The next day, the *Dublin Evening Post* explained Livingstone's inaudibility: 'His voice had suffered severely from constant speaking under trees, which had no covering but the vault of heaven, and he regretted that he was not able to make himself better heard'.[39] For Caroline Fox, having seen – and heard him – in Dublin in 1857, Livingstone's work was important for the hope he brought to Africa. 'But most one admires the earnest simplicity of the man, who always seemed as if he had so much rather be doing the work than talking about it.'[40] As Fox and others recognised, what was significant was not Livingstone's audibility but his celebrity. Thousands went to the 1864 Bath meeting where Livingstone spoke after his Zambesi Expedition and where the whole meeting of section E was coloured by the personal invective between Alfred Russel Wallace and John Crawfurd over their contrasting views on race, slavery and the treatment of the Maoris and the death by suicide of the African explorer John Hanning Speke on nearby Bathampton Downs (Figure 3.3).

On 16 August 1872, when Henry Morton Stanley spoke to over 3,000 people at Brighton, public excitement at seeing him was heightened further when Stanley departed from his prepared speech and inveighed against his detractors. The public audience applauded Stanley. His chairman, Francis Galton, rebuked the public cheering: 'I must beg to remind you that this is a serious society constituted for the purpose of dealing with geographical facts and not sensational stories.'[41] In these words Galton echoed his own presidential address wherein he had seen section E to be not a geographical society (that remit belonging to the Royal Geographical Society, of which he was a distinguished Fellow) but rather 'a constituent of a great scientific organization', a forum that should enable geographers to 'secure the attention of representatives of all branches of science'.[42] What Galton saw as an institutional space for the development of geography's scientific credentials, Stanley used – to the public's delight – as an opportunity for sensational narratives, personal invective and public acclaim as, at the same time, he used the press to promote his own glamorous self-image (Figure 3.4). Such things had happened before: giving a talk on western equatorial Africa to the 1861 Manchester meeting, the explorer Paul Du Chaillu only hinted at his work on the gorilla, but to prolonged applause and cries of 'Go on, go on' his audience – prompted, perhaps, by the excitement engendered by Darwin's *Origin of Species* two years earlier – demanded he elaborate on the topic to the neglect of his scientific focus.[43]

The science of geography

Figure 3.3 'View of the city of Bath, from the south-east' in 1864. The city was the scene of huge crowds to hear David Livingstone and of heated debates over human science in the BAAS. As the editorial to this picture noted, 'Bath undoubtedly offered an attractive *locale* for an assembly of the devotees of science. In the physical features of the neighbourhood it presents more than one problem well worthy of, but still awaiting, solution. It seems to challenge scientific investigation'

Experiencing geographical science depended then not simply upon differences between 'professionals' and 'amateurs', but between leading proponents of geography's still uncertain disciplinary status and celebrity practitioners whose renditions of exploration were rejected by the discipline's proponents yet eagerly anticipated and raucously received by public audiences. There was a strong sense, evident by the 1850s and satirised in the press in the 1880s, that geography was the 'ladies' section' (Figure 3.5). This schism between specialist knowledge and nascent disciplinary identity and popular public interest was heightened by the presence of leading explorers and the use of visualising technology ('magic lanterns' with their glass plate slides, and hanging wall maps) which lent an air of theatricality to section E's proceedings. This is not to state that specialist and technical papers always repelled audiences and that exploration narratives and illustrated talks on popular subjects always attracted. It is to suggest that what one took from section

Figure 3.4 Henry Morton Stanley, 'discoverer' of David Livingstone (*cf.* Livingstone's own views on Stanley's discovery of him in Figure 2.7 here)

E meetings depended not just upon urban location but upon which elements one attended in which sites, upon whom one heard (or not) and what one saw. As an anonymous commentator noted of the 1906 geography proceedings, 'A large concert hall was assigned to the section, which unfortunately had so bad an echo that it was very difficult to hear the speakers, and the discussions were curtailed as a consequence. With a crowded audience, the echo did not occur; but popular though the geographical section is, the whole hall was never filled during its meetings.'[44]

Figure 3.5 Apportioning responsibility for attending BAAS science. 'Paterfamilias, being unable to dismember himself to attend the various lectures, divides his family into eight sections, A to H. He will receive their reports on their return.' Note the clear association between geography, section E – as the father issues the ticket for that section – and attendance by women

Conclusion

To work from the place of geography in the BAAS to the workings of section E to the content of section E meetings and to presidential addresses and sectional papers is to work across institutional and epistemological boundaries and at different scales. What for Mill and Garnett was fond remembrance of institutional times past has been shown to be altogether more complex.

In the century from 1831, geography enjoyed a relatively low level of funding in respect of the mathematical, geological and physical sciences but was funded at greater levels than several other sciences and by a series of small grants and in association with other sciences (Table 3.1). Work of a geographical nature also appeared in other sections' programmes. As one contemporary remarked, 'it is well to remember that "Geography at the British Association" is a considerably wider title than "The Proceedings of Section E"'.[45] In broad terms, the subject underwent a shift in focus from what contemporaries understood as an exploratory tradition towards a scientific emphasis. Yet it did so differently in relation to certain themes within geography as a whole and differently within different sectors of the BAAS (*cf.* Tables 3.2 and 3.3).

Interpretations of the science of geography and the practices of section

E within the BAAS thus differ according to whether we look at grants awarded, presidential addresses, programmes of papers or audiences' reactions and practitioners' expectations. The demand for what contemporaries considered 'scientific geography' was evident in addresses from section E presidents as early as 1872 and was reiterated by later senior officers – with 'growing clarity' to paraphrase Mill – throughout the last quarter of the nineteenth century. In terms of grants for exploratory work, there was a relative decline from the early twentieth century and, in the papers given to public audiences at BAAS meetings, a sharper downturn in such work by 1900 if not before. But there was always a tension within the section, perhaps within the BAAS, over geography's place as a science: to do with geography's scientific credentials, its practitioner community, its content and its audiences. As was observed at the turn of the nineteenth century:

> Considered as a whole, the proceedings of Section E, while no less popular than in previous years, were more satisfactory on account of the higher average scientific value of the papers than is usually the case. It is, however, questionable whether the labour of organising such a meeting is not in great part thrown away. The number of geographical specialists in this country is small, and a regrettably large proportion of these do not consider it worth while to attend the association, hence the discussions are apt to fall into the hands of amateurs, or even of faddists, while the large audiences attracted by any paper which is supposed to be likely to turn out sensational gives to the public an utterly wrong impression of what geographical science is, and at the recent meeting led one newspaper to refer in good faith to the contrast between "the geographical and scientific sections".[46]

We must be careful, then, in how we approach geography's emergent complexity as a science and as a subject. And further implications follow. One concerns geography's place as a science within the BAAS in terms of the history of geography. Another centres upon geography's place as a science within the Association as a chapter in the civic history of British science. Rather than look for a defining 'moment' within the BAAS at which it is possible to discern the emergence of a 'modern' geography, the evidence here discussed has revealed a state of intellectual flux and, even, constant tension. What was (and was not) geography was never always strictly separate from other sciences in funding terms (Table 3.1) or in terms of associational activity. Geography's content and administration within section E, and section E's connections with other sciences, were always under negotiation. Geography was being more actively promoted within the BAAS and in diverse ways decades before figures such as J. S. Keltie and H. J. Mackinder (both active members of section E and leading figures in the RGS) campaigned in 1885 and 1886 for the presence of a 'new', modern, university geography geared towards educational improvement (see Chapter 5). Geography's institutional history does not equate with the subject's disciplinary history. There are always differences as to what geography meant, differences which may be recoverable from study of private correspondence rather more than from printed records: what at any one time was made public by the subject's leading proponents may have been far from what was being subscribed to by members of that institution pursuing geographical sensation.

This is to highlight the differences between geography's popular appeal – in BAAS talks with an 'exploratory tinge' – and the promotion of academic research which reveals an emergent thematic specialism, perhaps particularly after about 1900 and with respect to geomorphology, historical, social and regional geography. Section E negotiated its own scientific credibility – over what geography was, how it should be funded, which body if not the BAAS alone might best undertake the resolution of geographical problems and of geography's place in relation to other sciences. To consider the 'making' of geography as a science in terms of its production, intellectual status, financial support and public reception requires that we show more exactly how, when and where geography was being worked at and being worked out in its different institutional and intellectual spaces.

Notes

1 H. R. Mill, 'Geography at the British Association: a retrospect', *Scottish Geographical Magazine* 47 (1931), pp. 336–53, quote at p. 337.
2 Mill, 'Geography at the British Association', p. 344.
3 A. Garnett, 'IBG: the formative years – some reflections', *Transactions of the Institute British Geographers*, 8 (1983), pp. 27–35, quote at p. 29.
4 See, for example, O. J. R. Howarth, 'The centenary of section E (Geography)', *Advancement of Science*, 8 (1951), pp. 151–65, and S. Beaver, 'Geography in the British Association for the Advancement of Science', *Geographical Journal*, 148 (1982), pp. 173–81. In more recent work, the place of geography in the BAAS is briefly mentioned in R. J. Johnston, 'The institutionalization of geography as an academic discipline', in R. J. Johnston and M. Williams (eds), *A Century of British Geography* (London: British Academy in association with Oxford University Press, 2003), at p. 59, and R. J. P. Kain and C. Delano Smith, 'Geography displayed: maps and mapping', in Johnston and Williams (eds), *A Century of British Geography*, at pp. 386–7. The BAAS is overlooked by D. Livingstone, 'Geography', in R. C. Olby, G. N. Cantor, J. R. R. Christie and M. J. S. Hodge (eds), *Companion to*

the *History of Modern Science* (London: Routledge, 1990), pp. 743–60, by M.-C. Robic, 'Geography', in T. Porter and D. Ross (eds), *The Cambridge History of Science*, Vol. 7, *The Modern Social Sciences* (Cambridge: Cambridge University Press, 2003), pp. 379–90, and by C. W. J. Withers, 'A partial biography: the formalization and institutionalization of geography in Britain since 1887', in G. S. Dunbar (ed.), *Geography: Discipline, Profession and Subject since 1870: An International Survey* (Dordrecht: Kluwer, 2001), pp. 79–119.

5 J. Golinski, *Making Natural Knowledge: Constructivism and the History of Science* (Cambridge: Cambridge University Press, 1998), p. 6. For a fuller review, see D. Demeritt, 'What is the "social construction of nature"? A typology and sympathetic critique', *Progress in Human Geography*, 26 (2002), pp. 767–90.

6 The key role of geologists such as Roderick Murchison, Adam Sedgwick, George Greenough and others in the early development of the Association is clear in J. B. Morrell and A. Thackray, *Gentlemen of Science: Early Years of the British Association for the Advancement of Science* (Oxford: Clarendon Press, 1981), and J. B. Morrell and A. Thackray (eds), *Gentlemen of Science: Early Correspondence of the British Association for the Advancement of Science* (London: Royal Historical Society, 1984).

7 John Crawfurd, president of the Ethnological Society from 1861 and a regular speaker at RGS and BAAS section E meetings, was an advocate of close relations between these bodies. But from at least September 1861 he was arguing for a separate identity for ethnology within the BAAS: see British Library, Murchison Papers, Add. MS. 46125, fols 483–4, and Chapter 6 here.

8 On which point, see the essays in D. Cahan (ed.), *From Natural Philosophy to the Sciences: Writing the History of Nineteenth-Century Science* (Chicago: University of Chicago Press, 2003); M. Daunton (ed.), *The Organization of Knowledge in Victorian Britain*; Porter and Ross (eds), *The Cambridge History of Science*, Vol. 7, *The Modern Social Sciences*; B. Lightman (ed.), *Victorian Science in Context* (Chicago: University of Chicago Press, 1997).

9 O. J. R. Howarth, *The British Association: A Retrospect* (London: British Association, 1931), Appendix I, pp. 266–92.

10 M. Jones, 'Measuring the world', p. 329.

11 University of Oxford, School of Geography and Environment, Mackinder Papers MP/C/100, pp. 50, 51.

12 On which matters, see for example T. Gieryn, 'Boundary work and the demarcation of science from non-science: strains and interests in the professional ideologies of scientists', *American Sociological Review* 48 (1993), pp. 781–95; T. Gieryn, *Cultural Boundaries of Science: Credibility on the Line* (Chicago: University of Chicago Press, 1999); T. Guston, *Between Politics and Science: Assuring the Integrity and Productivity of Research* (Cambridge: Cambridge University Press, 2000).

13 With respect to work on the exploration of central Moab, for example,

Section E Minutes for 5 August 1871 record the granting of a further sum of £100, the initial grant 'having been found insufficient for the object': RGS, SSC/11 File 2, Minute Book of Section E, 1865–1873, p. 18. It is possible that the initial funding had been granted as early as 1865, since, at the Birmingham meeting that year, the Committee of Recommendations agreed upon £100 to Murchison, Rawlinson and Tristram 'to aid in the Geographical objects of [the] Palestine Exploration Fund': RGS, SSC/11 File 2, Minute Book of Section E, 1865–1873, p. 2, Friday 8 September 1865. It is also clear that this work was privately funded in addition to BAAS and RGS support: in a letter of 9 August 1871 Murchison wrote to Henry Yule to note, 'We have got another £100 for Moab exploration, i.e. in addition to £1000 drawn last year, & 3 private gentlemen will make this up to £500 for Messrs Palmer & Tristram to explore': British Library, Murchison Papers, Add. MS. 46128, fols 360–1. Some of this work was reported upon to the RGS: see C. W. Wilson, 'Recent surveys in Sinai and Palestine', *Journal of the Royal Geographical Society* 43 (1873), pp. 206–40. For an overview of this shared attention, see H. Goren, 'Sacred, but not surveyed: nineteenth-century surveys of Palestine', *Imago Mundi*, 54 (2002), pp. 87–110. Wilson, by then Director of the Ordnance Survey, addressed the Palestine Exploration Fund in 1899: C. W. Wilson, 'Address delivered at the annual meeting of the Fund', *Palestine Exploration Fund Quarterly Statement*, 1899 [no volume], pp. 304–16.

14 RGS, SSC/11 File 2, Minute Book of Section E, 1865–1873, p. 19, records thanks to Pandit Manphal, a Nepalese, 'for his service to Geographical Science'. This is exactly consistent with the reward given by the RGS to Nepalese and Tibetan natives in mapping projects and it is likely the two bodies collaborated in this work and drew upon the same local intermediaries: see K. Raj, 'When human travellers become instruments: the Indo-British exploration of Central Asia in the nineteenth century', in M.-N. Bourguet, C. Licoppe and H. O. Sibum (eds), *Instruments, Travel and Science: Itineraries of Precision from the Seventeenth to the Twentieth Centuries* (London: Routledge, 2003), pp. 156–88.

15 MS. Dep. BAAS 332, Correspondence and Papers of the Committee on the Teaching of Geography.

16 The 1856 'Niger deputation', consisting amongst others of Sir R. I. Murchison, Sir Henry Rawlinson, General Edward Sabine and Professor Owen, is outlined in *Report of the Twenty-sixth Meeting of the British Association for the Advancement of Science* (London: John Murray, 1856), p. xli. The 1873 recommendation on Arctic exploration was directed at the Prime Minister: 'In November last, Sir Bartle Frere, President of the Royal Geographical Society, requested Mr. Gladstone to receive a joint deputation from the Royal Society, the Royal Geographical Society, the British Association, and the Dundee Chamber of Commerce, on the subject of an Arctic Expedition. Mr. Gladstone declined to receive a deputation, but requested an application stating reasons, in a written form.' See *Report of*

the Forty-third Meeting of the British Association for the Advancement of Science (London: John Murray, 1874), p. xlviii.
17 C. W. Wilson, 'Recent surveys in Sinai and Palestine'; E. G. Ravenstein, 'On bathy-hypsographical maps; with special reference to a combination of the Ordnance and Admiralty charts', *Proceedings of the Royal Geographical Society and Monthly Record* 8 (1886), pp. 21–7.
18 A comment in the *Sheffield Daily Telegraph* of 21 August 1879 is echoed in other evidence: 'The Geographical section is always a popular one, because little or no antecedent scientific knowledge is necessary to enable the listeners to comprehend all the points in the memoirs read.' MS. Dep. BAAS 415.
19 [Anon.], 'Anthropology and the British Association', *Anthropological Review* 3 (1865), pp. 354–71.
20 Yorkshire Philosophical Society, Volume of news cuttings, 1881–1903 [supplement to *The Guardian*, 14 September 1881].
21 W. D. Cooley, 'Synopsis of a proposal respecting a physico-geographical survey of the British Islands, particularly in relation to agriculture', *Report of the Sixteenth Meeting of the British Association for the Advancement of Science* (London: John Murray, 1846), pp. 72–3. For the claim that 'Undoubtedly, this was a response both to the famine conditions then afflicting parts of Scotland and Ireland as well as to Robert Peel's repeal of the Corn Laws', see R. Bridges, 'William Desborough Cooley', in H. Lorimer and C. W. J. Withers (eds), *Geographers Biobibliographical Studies* 27 (2008), pp. 43–62, quote at p. 54.
22 Although the subject of geography was a central feature of some presidential addresses before 1858 (e.g. Sir Roderick Murchison in 1848), sectional relationships between geography and geology (section C) meant that section E addresses were not consistently given until 1858.
23 Sir J. Hooker, 'Transactions of the sections: Geography', *Report of the Fifty-first Meeting of the British Association for the Advancement of Science* (London: John Murray, 1882), pp. 727–38.
24 R. I. Murchison, 'Transactions of the sections: Geography and Ethnology', *Report of the Thirty-fourth Meeting of the British Association for the Advancement of Science* (London: John Murray, 1865), pp. 130–5.
25 F. Galton, 'Transactions of the sections: Geography', *Report of the Forty-second Meeting of the British Association for the Advancement of Science* (London: John Murray, 1873), pp. 198–202.
26 Lieutenant General. R. Strachey, 'Transactions of the sections: Geography', *Report of the Forty-fifth Meeting of the British Association for the Advancement of Science* (London: John Murray, 1876), pp. 180–8.
27 Strachey, 'Transactions of the sections: Geography', p. 184.
28 RGS-IBG Archives, Correspondence Block CB 1871–1880, [A. Keith] Johnston to [H. W.] Bates, 30 November 1875.
29 C. R. Markham, 'Transactions of the sections: Geography', *Report of the Forty-ninth Meeting of the British Association for the Advancement of Science* (London: John Murray, 1879), pp. 420–32.

30 RGS-IBG Archives, C. R. Markham Correspondence Block CB 1871–1880, and Markham Collection 47, 'The Royal Geographical Society', p. 429.
31 G. T. Goldie, 'Transactions of the sections: Geography', *Report of the Sixty-seventh Meeting of the British Association for the Advancement of Science* (London: John Murray, 1907), pp. 611–17.
32 Mill, 'Report to Council', p. 524.
33 J. M. MacKenzie, 'The provincial geographical societies in Britain, 1884–1914', in M.Bell, R. Butlin and M. Heffernan (eds), *Geography and Imperialism, 1820–1940* (Manchester: Manchester University Press, 1995), pp. 93–124.
34 H. R. Mill, 'Report to Council on the British Association meeting at Newcastle, 1889', *Scottish Geographical Magazine* 5 (1889), pp. 606–8, quotes at p. 606.
35 W. S. Dalgleish, 'Geography at the British Association, Oxford, August 1894', *Scottish Geographical Magazine* 10, pp. 463–73, quote at p. 463.
36 MS. Dep. BAAS 147, Monday 2 July 1860.
37 In the two-year period 1889–1891, for example, ninety-eight papers given at the BAAS Section E (Geography) sessions were published in other societies' journals: forty-nine by the Royal Scottish Geographical Society, thirty-five by the Manchester Geographical Society, three by the Tyneside Geographical Society, three by the Glasgow Philosophical Society, two each by the Belfast Natural History and Philosophical Society and the Birmingham Natural History and Microscopical Society and single papers by the Essex Field Club, the Leeds Geological Association, the Somersetshire Archaeological and Natural History Society and the Warwickshire Naturalists' and Archaeologists' Field Club. The Corresponding Societies of the BAAS (formally begun in 1884) played an important role in this: on the Corresponding Societies, see R. M. MacLeod, J. R. Friday and C. Gregor, *The Corresponding Societies of the British Association for the Advancement of Science, 1883–1929: A Survey of Historical Records, Archives and Publications* (London: Mansell, 1975).
38 H. R. Mill, 'Report to Council', *Scottish Geographical Magazine* 3 (1887), pp. 521–30, quotes at pages 525, 527, 529.
39 *Dublin Evening Post*, 1 September 1857.
40 H. N. Pym (ed.), *Memories of old Friends, being Extracts from the Journals and Letters of Caroline Fox* (London: Smith Elder, 1882), p. 314.
41 T. Jeal, *Stanley: The Impossible Life of Africa's Greatest Explorer* (London: Faber, 2007), p. 140.
42 Galton, 'Transactions of the sections: Geography', p. 198.
43 *Manchester Courier and Lancashire General Advertiser*, 7 September 1861, p. 7. On the wider context of Du Chaillu, see J. Mandelstam, 'Du Chaillu's stuffed gorillas and the savants from the British Museum', *Notes and Records of the Royal Society of London* 48 (1994), pp. 227–45.
44 [Anon.], 'Geography at the British Association', *Geographical Journal* 28 (1906), p. 496.

45 [Anon.], 'Geography at the British Association, Ipswich, 1895', *Geographical Journal*, 6 (1895), pp. 460–5, quote at p. 465.
46 [Anon.], 'Geography at the British Association, Bristol, 1898', *Geographical Journal* 12 (1898), pp. 377–85, quote at p. 385.

4

The dominion of science and geographies of empire: the BAAS overseas, 1884–1929

Looking back on the 1905 South African meeting of the British Association, Professor J. Y. Simpson recalled his experiences thus:

> To tell the story of the British Association in South Africa is to furnish an itinerary of nearly twenty thousand miles of wandering, and to furnish *résumés* of numerous papers of first-rate importance, dealing either with local problems of general interest or with the most recent advances in modern science. It is also to recount the lighter side of the most gigantic picnic ever organised, and unstintingly to shower praise on many officials at home and in the colonies who worked unsparingly to make the visit the unqualified success it ultimately proved to be.[1]

Securing connections between scientists at home and in the empire had been a concern of the BAAS at its foundation: the object 'to promote the intercourse of those who cultivate Science in different parts of the British Empire with one another and with foreign philosophers' speaking directly to this intention'.[2] The 1905 South Africa meeting was the third such overseas meeting, earlier ones having been in Montreal in 1884 and in Toronto in 1897, and itself preceded those in Winnipeg in 1909, Australia in 1914, in Toronto again in 1924 and in South Africa in 1929.

As Simpson's testimony suggests, BAAS overseas meetings sought to use science as a route to knowledge of, and material improvement in Britain's dominions. Quite how they did so – whether as 'gigantic picnic', through papers of 'first-rate importance' yet differing purpose and from concerns to bond Britain's empire through scientific endeavour – is the subject of this chapter. What follows is less concerned with individual meetings or with national settings *per se* than it is with the themes that connect them, namely the imperial context to BAAS science and the work done in section E meetings overseas. In relation to what we have seen about the nature of BAAS meetings and the workings of section E, this chapter considers how foundational objectives concerning science and empire worked in practice, and how and where geography as a subject and a section fitted within the science–empire nexus. On the one hand, addressing the historical geography of associational science as a form of empire making by examining the different sites and analytic scales through which the BAAS and section E operated in these imperial venues is to disclose the historical geography of BAAS science at different geographical scales. On the other, it is to see how the science of geography worked in and was shaped by particular colonial settings.

Science and the British Empire: the imperial agenda of the BAAS

How are we to approach an understanding of the internationalism or, perhaps better, the internationalisation of science? One theme common to study of the connections between science as a transnational form of institutionalised knowledge making and the power of the state has been the view of scientific institutions – in Britain, the Royal Botanical Gardens at Kew and the Royal Geographical Society, for example – as metropolitan centres of information, co-ordinating spaces for knowledge's global classification and of the 'globalising' expeditionary cultures which underlay the collection and display of 'others' to 'home' audiences. In such a sense, science, made locally yet dis-located from that setting, became global by virtue of the co-ordinating role of one or more institutions in given national contexts having international 'reach'.[3] A further and related theme, characteristic of eighteenth-century Europe and earlier, is that 'cosmopolitanism' of intellectual exchange in which individual natural philosophers and men of letters wrote to one another, or in which learned societies had formal ties, even as their respective nations' political interests may have been at loggerheads: shared interest in human knowledge, it was argued, would overcome the friction of geographical distance and political dissent.[4] During the nineteenth century, and as science became increasingly professionalised, specialised and institutionalised, so national associations of learning drew upon international corresponding members and, from the later 1800s especially, subject-specific international collaboration became a feature of science's co-operative operation over and above national boundaries even as it also served national interests.[5]

It is difficult to know quite how the BAAS in its overseas meetings fits into these interpretative models. Relatively little work has been done on the Association's internationalist agenda. For Guiseppi Pancaldi, the Association's 'scientific internationalism' evident in its objectives at foundation was apparent in three ways: in early plans, led by Charles Babbage, for a migratory European scientific academy (never realised);

in the co-operation between nations from the mid 1830s in studies of terrestrial magnetism; and because the BAAS itself provided a 'model' that could be exported – Roy MacLeod has shown that it was – to promote scientific activity in other countries.[6] For Pancaldi, the scientific internationalism proclaimed by the British Association and other parallel bodies helped 'sustain the conception of science as a cosmopolitan undertaking, and has promoted the elaboration of concrete plans for scientific co-operation'.[7] Seen thus, the BAAS was an agency for the promotion of science as a transnational intellectual enterprise: science was without frontiers, the BAAS an agent of shared intellectual endeavour unhindered by politics and not shaped by geographical circumstance.

For Michael Worboys, connections between the British Association and the British Empire were apparent in the first fifty or so years of its activities in scientific reports on the colonies and in shared initiatives over terrestrial magnetism, but the 'science–empire' connection was most evident in the period 1880–1940, especially between 1914 and 1940. This was because of a shared reformist sense that science could aid the empire, an agenda Worboys terms 'social imperialism'.[8] From consideration of five of the Association's overseas meetings (excluding South Africa in 1905 and 1929) Worboys has shown that, in Britain, the alliance of science with imperialism added weight to claims about the utility of science as a form of imperial politics. Simply, the Association had a key role in promoting through science a notion of what the empire might become – a new 'Dominion' or 'Commonwealth of Science'. Worboys' claims are echoed in Saul Dubow's study of the South African meetings. Dubow shows how the 1905 meeting was framed not just by notions of political reparation in the wake of the Boer Wars but also how the later meeting, in 1929, was planned to give greater emphasis to the science in and of Africa by colonial scientists rather than by, as earlier, visiting experts.[9]

As a body established to be peripatetic, the BAAS's overseas meetings can be seen as a natural consequence of its foundational commitment and were certainly part of the internationalisation of individual scientific exchange and of subject-specific scientific societies from the later nineteenth century. Yet the BAAS overseas was, I suggest, interpreted differently at different times and in different places: colonial scientists and political figures and popular audiences overseas welcomed BAAS scientific delegates, in the early meetings especially, as visiting experts and agents of imperial authority. Science in the colonies was vital to colonial political development. The coming of the BAAS represented an opportunity to present new sites and sights for imperial visitors and, in turn, have knowledge of colonial circumstances travel back to Britain. The BAAS also co-ordinated various imperial geographical projects from its London-based offices, although the personnel who did them came from universities in the United Kingdom and elsewhere. Both by being 'out there' and in the connections that followed, British scientists and geographers learned from exposure to colonial environments and through the workings of 'dominion science' as, locally, they helped make science and geography serve the needs of empire. But for BAAS meetings to be successful overseas they had in several senses to overcome the facts of geography – each of the overseas meetings required oceanic voyaging, many were orchestrated around long train journeys – and to negotiate a balance between 'local' papers and excursions and broader topics as well as allow new findings to travel safely back to Britain. To BAAS figures credibility was on the line: theirs in being distant experts but in new locale, a transient status as colonial scientists alongside indigenous practitioners keen to show their worth.

Associating overseas

The decision to have the annual meeting overseas, in Montreal in 1884, followed an invitation from the then Canadian Governor General, the Marquess of Lorne, who, with others, argued that its coming would strengthen Canada's place in the British Empire. Political motives were paralleled by economic incentives: from the Ottawa government towards British scientists' expenses, and from shipping and railway companies for transatlantic travel and the excursions. The Royal Society of Canada, founded in 1882, wanted to attract the BAAS in order to boost Canadian science, locally and in international context. To this end, invitations to the 1884 meeting – and more strongly in 1897 in Toronto – were extended to numerous leading American scientists (which accounts for the relative increase in 'foreign delegates' in these years: see Table 2.1). Finally, the citizens of Montreal saw local civic value in having the BAAS come, especially then given the loss from their city of Canada's Geological Survey and the financial difficulties faced by McGill University.[10]

But while Montreal represented a key opportunity for the BAAS as 'an agent of British scientific diplomacy' at a time when its fortunes within Britain were in relative decline,[11] the Association's leading figures were not initially agreed over the need for and the presumed benefits of overseas meetings. This is clear in letters to her mother, printed privately, from Clara Strutt, Lady Rayleigh, wife of the then BAAS president:

> When the decision was come to at Southampton [the 1882 meeting] to hold the meeting of 1884 in Canada there was widely expressed disapproval of

the step, and doubt as to its legitimacy; but the prospect of entertaining the upper thousand of English [sic] science has evidently so greatly gratified our Canadian brothers that even the most stiff-necked opponent of the migration must be compelled to give in if he has a shred of good nature and brotherly feeling left.

As she further noted, going overseas could extend the same function within the empire as the BAAS had done in Britain:

There are doubtless a few grumblers who will maintain that the Montreal assembly will not be a meeting of the *British* Association; but after all this Imperial Parliament of Science could not be better occupied than in doing something to promote science in one of the most important sections of the British dominions. Indeed, since some maintain that so far as this country is concerned it has almost ceased to have a *raison d'être*, might it not extend its functions and endeavour to exercise the same effective influence on the promotion of science in other parts of the Empire as it has undoubtedly done in the past in the Mother Country?[12]

This expression of the Association's conjoint role in imperial politics, the promotion of science in local and international settings and science's civic significance as a form of imperial utility is evident to varying degrees in each of the overseas meetings. Overseas meetings were supported by scientific bodies and civic and political authorities in the same way as we have seen for the British meetings: writing in 1902 to extend the formal invitation to South Africa in 1905, for example, Sir David Gill did so on behalf of the governments of the Cape Colony and Transvaal, the South African Philosophical Society and the South African Association for the Advancement of Science, the mayors and corporations of Cape Town, Johannesburg, the chambers of commerce of those towns and Johannesburg's chamber of mines and its stock exchange.[13] In his speech of welcome in 1884, Canada's Governor General emphasised the connections between the reach of science and of empire as above yet linking national interests:

In the domain of science there can be no conflict of local and imperial interests – no constitution to revise – no embarrassing considerations of foreign and domestic policy. We are all partners and co-heirs of a great empire, and we may work side by side without misgiving, and with a certainty that every addition to the common fund of knowledge and mutual enlightenment is an unmixed advantage to the whole empire.[14]

Petitioners within Canada for a second meeting stressed the significance to Canada of the 1884 meeting, noting that it had had 'a marked influence upon the intellectual growth of the Dominion through the manifest advantages derived from intercourse and association with the eminent and learned men who compose that Association':

> Canada has reaped substantial advantages from the results of the visit to Montreal, in 1884, of the persons, who, prominent in numerous branches of political and economical science composed the several sections of the Association, who carried back with them lively recollections of our social, moral and industrial standing. The Dominion became better known in England, and we now enjoy the fruits in the important delegations of agriculturalists, and others engaged in manufacturing and mercantile pursuits, who visit us from year to year.[15]

A diffusionist rhetoric for science – for the common good, not empire's benefit – was also evident in the socialist periodical *Justice*, which carried notice of the 1884 Montreal meeting.[16]

In South Africa, the 1905 meeting was seen as a means of repairing unitary notions of empire and nation in the wake of the Boer Wars, of promoting the South African Association for the Advancement of Science, which had been founded in 1902 based on the BAAS and, from direct observation and first-hand encounter of a still youthful country, of dispelling established myths concerning South African affairs[17] (Figure 4.1). Reviewing and comparing the Canadian visits and the 1905 South African meeting, the *Times* considered it

> unnecessary to dwell at length on the Imperial aspect of intercourse of this character between the mother country and the Colonies. . . . The aims and work of the British association are essentially scientific, and as such are cosmopolitan. At the same time it may unhesitatingly be affirmed that the promoters of these occasional migrations of the Association would not attain the full realizations of their hopes if the visits to the Colonies did not in some measure further the cause of imperial unity.[18]

By 1929, the focus of the BAAS meeting was to promote the 'Africanisation' of South African science, by emphasising local work and by placing South African science and scientists within international not just dominion context. Science, as the Mayor of Cape Town stressed in welcoming delegates in July 1929, was essential to the march of civilisation. South Africa was determined to enjoy, with others, 'the countless advantages' accruing from 'the efforts and sacrifices made by the pioneers of thought, invention and industry'.[19]

In Australia, plans begun as early as 1910 to bring the BAAS there in 1914 stressed the benefits for that country, for science and for the British Empire: 'It would stimulate Imperial interest,' noted the Lord Mayor of Sydney, 'and would send back every visiting delegate as a zealous missioner, well informed, and anxious to help in peopling and effectively

Figure 4.1 Science for a youthful empire. The cover of the programme of the South African Association for the Advancement of Science in collaboration with the BAAS, 1905

occupying the vast continent of Australia.' Having the BAAS come 'could do nothing but good to the Commonwealth in making known its natural resources and potential wealth' and would reveal 'to the world at large the wonders and beauties' of Australia. Further, 'the more people we can get to visit Australia the better for Australia, and I venture to say the better for Empire'.[20] Civic figures were of like mind: writing to Melbourne's lord mayor in June 1910, the Lord Mayor of Sydney spoke of the 'the great advantages that the Commonwealth must reap in so many ways from the proposed visit. With her vast natural resources Australia has untold fields awaiting proper scientific development, and the impetus the visit must give to scientific method and research is at present much needed.'[21] To the editor of *The Age*, 'The chief advantage to be gained from such a visit as that proposed would be the strong appeal made to the popular imagination in this country concerning the potency of science in modern life. . . . The meeting is like a brief annual session of a British parliament of science. . . . As hosts of the British Association, the Australian people would be recognised all over the world as a nation taking a serious interest in the deeper problems of humanity'.[22]

These general intentions to secure imperial networks of information exchange through the work of the BAAS depended upon managing the meetings in certain ways. This included selecting the best people to ensure these intentions would be realised: as W. A. Herdman the British oceanographer put it to A. D. C. Rivett, the organising secretary for the 1914 Australian meeting:

> It must be remembered that some of the famous men you refer to are no longer the most active workers in Science – They are not now in the fighting line & if they did go, would not perhaps be able to do much. The next set – such men as are Presidents of Sections are those who are now in the forefront & whose names if not yet known to the man in the street will be soon. Could this not be put judiciously at the right time & in the right way before the Australian public – with perhaps a short notice of the work of each of the leading men in our party?[23]

The imperial agenda articulated by the civic figures leading invitations to the BAAS was rooted in concerns that the visitation of the BAAS should benefit the host country and that the visit of BAAS scientists and the conduct and nature of Association science should also benefit Britain. As petitioners in Canada in 1893 made clear, Association science in Canada and imperial unity was about the material development of Canada and about informing 'metropolitan' authorities in Britain:

> Your petitioners beg further to represent that no better plan of making known throughout England the advantages of Canada as a home for the

emigrant, and as a field for the employment of capital can be conceived than the holding out to members of the British Association inducements to visit us. The British Association is divided into sections for the study of special subjects, such as mining, metallurgy, agriculture and social science, and many of the persons closely connected with each of these sections are eminently fitted to distribute, when they return home, that class of knowledge of the Dominion which should be available to intending settlers.[24]

R. L. Richardson, the reporter who travelled with the BAAS party on its 1884 post-meeting trans-Canada excursion, was of the view that 'The most important result of the visit will not be the impetus given to the study of science in the Dominion, although all admit the great importance of that result, but rather the opportunity afforded to Canada of demonstrating to the leading and thinking men of England, that with her great natural resources, she is the most desirable field for immigration now competing for the great surplus population of Britain'. In reply to a speech of welcome made in Calgary, Sir Richard Temple spoke of BAAS delegates being 'astonished at what they had seen' as they 'saw a vast prairie being converted into a paradise' and how these visitors 'would not forget to proclaim on their return to England what a grand and glorious country this was'.[25]

In sum, whilst the overseas meetings of the 'Imperial Parliament of Science' were about extending to the colonies the benefits of science as a civic and imperial cultural resource, they were seen by colonial authorities as a means to promote local improvement by having BAAS delegates and leading scientists return to Britain as political-scientific emissaries to spread the word on the economic opportunities as well as the scientific work being done. This was, in part, about testing what Knight has called the 'tyrannies of distance in British science', meaning that sense that London knew best and that the further away one got from the metropolis, the less able were the practitioners.[26]

Geography for empire, geographies of empire: the overseas work of section E

BAAS overseas meetings followed the established form: a Presidential address, sectional papers preceded by that section's presidential talk, social events and excursions and programmes of working men's lectures or other targeted activity. As in Britain and Ireland, BAAS meetings could, effectively, take over many of the civic spaces in a town or host institution. Differences in the overseas meetings centred upon the attention given to local colonial science as it was used to inform BAAS delegates in order that they and the sciences could travel back to Britain

Table 4.1   Date of publication and title, BAAS handbooks for overseas meetings, 1884–1929

Year of publication	Title
1884	*Handbook for the Dominion of Canada*
1897	*Handbook of Canada*
1905	*A Handbook of Capetown and Suburbs*
	A Guide to the Transvaal;
	Science in South Africa: A Handbook and Review
1909	*A Handbook to the Province of Winnipeg and the Province of Manitoba*
1914	*Federal Handbook*
	Handbook and Guide to Western Australia
	Handbook for New South Wales
	Handbook to Victoria
	Handbook of South Australia
	Tasmanian Handbook
1924	*Handbook to Canada*
1929	*South Africa and Science: A Handbook*

Source: BAAS Archives, Bodleian Library.

as evidence of colonial need, imperial opportunity and the credibility of colonial practitioners. These differences are apparent in the nature of the sectional programmes where concern was taken to incorporate themes of local importance. Excursions were larger in scope than their domestic counterparts and provided an important and mobile means to attain scientific knowledge. The handbooks for the overseas meetings drew upon established practice in stressing the scientific importance of given sites and, additionally, acted as guides to dominion geographies in providing textual space for the demonstration of colonial scientific expertise.

Publicising empire: BAAS handbooks

BAAS handbooks for the overseas meetings promoted both the dominion in question and the town or towns hosting the BAAS (Table 4.1). In South Africa and Australia, because meetings were, effectively, tours on the move, the guide books presented the activities and things to be seen as a sort of scientific timetable or sites. Although BAAS meetings in Canada were held, like those in Britain, in single towns, three of the four Canadian handbooks emphasised the connections between the nation as a whole and the dominion of science (the exception being the 1909 handbook for Winnipeg and Manitoba). The *Handbook for the Dominion of Canada* (1884) framed the first BAAS visit as an opportunity for British scientists to see a new country under something like laboratory conditions – a new and emergent geography being distilled in spirit:

> An Englishman [*sic*] visiting Canada for the first time . . . will see a people of his own race and language who have adapted themselves to a totally new set of conditions.
>
> In Canada, he will see the youngest nation in the world; and he will be able to converse with those who assisted at its birth. He will see institutions in process of construction, and a people, without a leisure class, busy working at them.[27]

Each volume adopted a common presentational and rhetorical style: a general review of the dominion in question, followed by thematic chapters written by colonial men of science. Maps – of climate, railways, topography and so on – helped present unknown geographies to outsiders: designed, as the 1897 handbook had it, 'to give for the use of scientific visitors to Canada a compact and systematic account of those features of the Dominion in which they might be presumed to have a general interest'.[28] Separate maps showing the climatic regions and seasonal variability in climate in the colonies were even prepared in advance of the meetings and distributed to delegates on board ship.

Credibility in the authorship of the handbook essays was vital for the texts to be taken seriously by BAAS scientific visitors: 'the Editors have endeavoured to get information at first hand from people who are known to be authorities with regard to the localities, or the branches of science that they describe'.[29] Griffith Taylor's essay 'The physical and general geography of Australia', for example, in the 1914 *Federal Handbook* detailed Australia's geography by natural regions, making clear to visitors and Australians alike the connections between climate, physiography and Australian industrial development.[30]

Even where distinguished colonial authorities offered summary guidance, a common discursive claim was the novelty of science in the dominions, the vital role that British science played in colonial development and the value that dominion men of science placed on the BAAS as an instrument of scientific and imperial advance. In his Introduction to the 1905 *Science in South Africa*, Sir David Gill, President of the South African Association for the Advancement of Science, noted how 'Few countries owe more to science than does South Africa'. He emphasised how, until recently, science there had depended upon incomers, but now that a 'few sons of the soil' had been trained by such men, 'there is evidence of a marked increase in true scientific work, and a hopeful prospect of more'.[31] Others agreed: where astronomy in the colony was strong, marine biology, meteorology and economic botany were particularly wanting.[32] His and others' view that science would aid the infant colony was evident in the programme for the joint meeting with the BAAS (see Figure 4.1). For Canada in 1924, the noted emphasis on geographical opportunity and material development was apparent in an accompanying handbook on Ontario (Figure 4.2).

BAAS overseas handbooks thus sought colonial self-promotion through science and did so by careful management of content even as they looked to BAAS visitors as discerning experts. They were not always, however, the result of colonial unity. The production in South Africa in 1905 and in Australia in 1914 of provincial or state handbooks in addition to national handbooks and town guides (see Table 4.1) in part reflected plans for given meetings: the 1914 *Handbook and Guide to Western Australia* was produced only for an advance party of BAAS delegates (the main conference never went there) who attended in order to see Australia's 'youngest State in development' – a visit received with 'wide-spread feelings of gratification'.[33] In part also, they were the consequence of competition within and between colonial, city and state authorities and different institutions over attracting the BAAS. The fact that the Royal Geographical Society of Queensland – which was otherwise much involved in regional description and the promotion of imperial themes locally – was not called upon to assist in a Queensland guide or in the federal handbook was a source of some acrimony. Elsewhere, pleas for standardisation of approach faltered in the face of personal rivalries and disputes over which subjects (such as mineralogy, economic botany) merited lengthier treatment as the more important sciences of empire.[34] Butlin has shown how historical geographical texts of the British empire in this period promoted themes of empire and environmentalism – including work by the colonial administrator and author Sir Charles Lucas, whose idea it was to have the 1914 BAAS meeting in Australia.[35] BAAS overseas handbooks similarly helped articulate notions of colonial dependency and geographical opportunity even as they sought to demonstrate scientific capacity with the aim, ultimately, of parity of status with their metropolitan visitors.

Debating science and empire: the programmes for section E and other sciences

A feature crucial to the content and success of section E paper sessions overseas and, thus, to the realisation of geography's place as a science of imperial utility, was the balance between papers of local interest and those of different purpose. This was a constant concern for local organisers. For that reason, Dubow's observation about the progressive 'Africanisation'

Figure 4.2 Science and national enterprise. The cover of the Ontario handbook issued in collaboration with the BAAS meeting, Toronto, 1924

of scientific topics between the 1905 and 1929 South Africa meetings may be tested with respect to the national focus afforded to science and geography in Canada between 1884 and 1924 and in Australia in 1914.

His claim is borne out from exchanges between officials planning the 1905 meeting. In a 'report on papers already promised for the B. A. meeting', it was noted that geography 'is perhaps the most difficult Section to get papers for, and with the exception of the above [a paper on 'Geographical notes on South Africa, South of the Limpopo' by F. S. Watermeyer] it is questionable whether we shall get any papers'.[36] They did, but largely from visiting scientists. The Oxford-based geographer A. J. Herbertson was much involved with the 1905 meeting. He afterwards wrote to John Scott Keltie in London about the daily running of the meeting and what any follow-up to it might involve:

> The section work went fairly smoothly; At Cape Town the new Municipal Buildings were hardly finished, & every day chairs had to be brought from the great hall to our section Room & taken back after our meetings. The passages were somewhat confusing and direction labels almost invisible or wanting. By 10.15 on the first day I had remedied this for Section E, for that it did not explain the very small attendance. Neither there nor here [Cape Town, and Johannesburg] have we had more than 40 or 50 in the room and 12 to 25 was a commoner audience. The proceedings were not especially thrilling, but some good discussions were raised even with small audiences – especially glaciation with Section C, Hutchins paper on Indigenous Forests, Trevor's paper on the Transvaal, Davis on the Geographical cycle in arid areas. I have retained the papers & hope you will find room for some in the G.J. . . . I have suggested research work here to several people & hope to get two or three papers for the research committee . . . . All recommendations whether with or without grants were accepted & not a 1/2d taken off any. We applied for reappointment of old committees, except terrestrial waves, and a new one on quantifying & composition of rainfall and run-off jointly with B, C, and G for wh. £10 was asked.[37]

For the reporter on the *Morning Post*, the 1905 meeting was of mutual benefit: 'In conversation and in the excursions about the country the British Association members picked up all sorts of useful ideas which will doubtless be worked out in monographs during the coming winter. On the other hand scientific research in South Africa, already well advanced, has undoubtedly received a useful stimulus'.[38]

There were also features distinctive to each meeting: in the personnel involved and in the focus of sessions organised jointly between section E and other BAAS sections. At Toronto in 1897, the meeting involved 'a number of the eminent geographers of the United States' and 'the communication of valuable reports from what for the moment may be termed the geographical departments of the Governments of Canada and the United States' (but the non-participation of Canada's only geographical society, the (French-speaking) Geographical Society of Quebec). Most speakers to section E were British geographers and/or foreign men of science: Hugh Mill, Vaughan Cornish, Ernst Ravenstein to name a few. Only three of the thirty-three papers presented in section E over its five-day programme in 1897 concerned Canada (each was given by a Canadian scientist). Although four papers on aspects of Canadian physical geography were given in association with sections A and C (mathematics and physics, and geology), and three on Canadian ethnography in association with section H (anthropology), the programme was dominated by visiting experts.39

By the 1909 Winnipeg meeting, the majority of papers referred to North America, mainly to Canada, and had been arranged as such in advance: foreign geographers such as De Martonne from France were also invited as were geographers in the United States.[40] The papers about Canada were of three sorts: agricultural geography, physical geography, in which, on the day-long excursions, Canada provided new sites and features of a scale bigger than most Europeans had encountered, and ethnography and ethnology (notably about Canada's native peoples). Section E papers on wheat at Winnipeg were lead by Albert Perry Brigham, secretary to the Association of American Geographers. Brigham saw the USA as the main producer of wheat in the Americas and argued that world wheat prices might be brought down if Canada continued to extend its wheat-growing areas. In response, a paper by Professor James Mavor of the University of Toronto on the likely underestimation of Canada's wheat cultivation between 1904 and 1909 led to animated discussions between them, Professor Chapman, president of section F (economics), Major Craigie, chairman of section K (botany), William Saunders, Director of the Dominion Experimental Farms and Dr Robertson of the Macdonald Institute in Montreal.

The geography of agricultural science was for these men a topic of debate across international borders and a form of imperial knowledge hotly debated by members of the geography, economics and botany sections. It also mattered locally, to the citizens of Winnipeg – which had been chosen as the site for the 1909 meeting in order 'to spread scientific knowledge and also for the purpose of aiding in the development of newer Canada'.[41] Whilst the BAAS and Canadian provincial authorities chose Winnipeg because visitors were exposed there, through science, to 'the real spirit of the west . . . its contagious optimism and self-reliant confidence', Winnipegers went chiefly to the wheat papers in order to get most benefit from such science as spoke to local circumstances.[42]

In Australia in 1914, printed reports note that 'the majority of papers in section E dealt with Australian geography'.[43] Private correspondence reveals the negotiations necessary to make this work and that such management varied by science according to organisers' views over the utility of different subjects. Orme Masson, of the University of Melbourne and Chair of the Federal Council for the 1914 meeting, was keen to ensure geography's parity with other subjects: 'We note the one third rule with extension to one half in Zoology, Geology, Botany, and Agriculture. The opinion was expressed that, acting on the same obvious principle, there should be a similar extension in Anthropology; and I myself would add, though it was not thought of then, in Geography. It must be remembered that Geography will include Antarctic work'.[44]

From the BAAS perspective, planning distant science with unknown people was problematic. 'Of course we must trust the local people far more than usual' [wrote O. J. R. Howarth to W. A. Herdman]. 'The worst thing is that neither Rivett nor I are much informed with the Sydney locals: they don't seem a very practical lot; we have done what we can to inspire them'.[45] Moreover, sections' presidential addresses were tailored to the presence in different Australian cities of relevant institutions and appropriate audiences: as Rivett responded to Masson,

> Physics and Chemistry should come where Schools are well established, in order to ensure some sort of professional local audience. Geology undoubtedly should be in Sydney. . . . Engineering seems right for Sydney, especially in view of the big bridge project. Martin would go to Brisbane to oblige us, but this would be a pity because his address on the white man's capacity for labour in the Tropics is intended by his Sectional Committee [Anthropology] to be the opening to a very big discussion on the subject, all arranged for the particular benefit of Australia.[46]

This evidence that the sectional sessions of different sciences reflected particular local circumstances, and that the connections between science and empire were worked out in different ways within an overall rhetoric of the Association's stimulus to colonial development, is supported by the section E programme in South Africa in 1929 and by plans to have there, as in other meetings, public lectures which put scientific matters across in an accessible way. In the 1914, 1924 and 1929 meetings, many such public lectures – the term 'Citizens' Lectures' was adopted – were provided by BAAS delegates in order that colonial science might be promoted by visiting experts.[47] Public lectures and even presidential addresses were used to legitimate dominion politics. When the BAAS visited Durban in Natal in 1929, for example, civic dignitaries and local newspapers used the visit to campaign for educational provision there, given that 'In Natal there was a different topography, a different geography, a different soil and a different spirit from the rest of the Union.'[48] In the 1929 section E sessions held in Cape Town and Johannesburg, local scientists and topics dominated proceedings. Collaboration between different sections – with section A on geodesy and hydrology, with section L (education) on the teaching of geography in South Africa and in England – further ensured that BAAS scientists spoke, simultaneously, to the worth of colonial geographies as scientific subjects in their own right and to the place of science in colonial development.[49]

As we have seen, geographical investigation of Britain's colonies took place in BAAS meetings within Britain, by research awards in aid of given topics such as that granted to Ernst Ravenstein in 1892 on the 'climatology and hydrography of tropical Africa' and to Sir Charles Lucas in 1926 on 'Geographical knowledge of tropical Africa' (Table 3.2), and through committees such as that on the Human Geography of Tropical Africa. This, under Lucas's chairmanship, undertook and directed work on population, environment and colonial development, reporting in detail at the 1929 South Africa meeting.[50] Other committees were established, or mooted at least, given the particular context to overseas meetings: joint plans by the geography and geology sections in Australia in 1914 (initially brought forward by the Governor of Fiji) to establish a series of benchmarks by which, at ten-year intervals, sea-level change affecting the islands in the Coral Sea might be measured came to nothing in the face of inaction from the Secretary of State for the Colonies.[51] Some committees – such as that on Soil Resources of the Empire (in existence between 1927 and 1936) and the Geodetic Committee (1917–45) worked in association with other sections and continued after the last overseas meeting. Their concern, as particularly with the Committee on Geography in Dominion Universities which was led by Cyril Fawcett and L. Dudley Stamp between 1931 and 1933, was with enhancing geography in the colonies as a form of imperial science.[52] Earlier, discussions about the ethnological survey of Canada, which began in 1884 at the Montreal meeting, moved back and forth across the Atlantic, being developed further in Toronto in 1897, and became a major topic of discussion in meetings in Dover, Bradford, Glasgow and Belfast between 1899 and 1902.[53] After World War I, discussions at Bournemouth on joint ethnological work between section H and section E turned in the Australian context to the question of that 'geographical and anthropological material which the Association understands was left by the Germans in New Guinea'.[54] In such ways, colonial science circulated between the BAAS overseas meetings as well as being made locally in and out of the colonies.

Presidential addresses overseas

In contrast to the localist agenda increasingly pursued in section E paper sessions overseas, only one of the geographical presidential addresses delivered there took the geography of that colony as its central motif. Most speakers concerned themselves with general remarks about the progress of geography, or looked to exploration and advances in topographic mapping as the big achievements of the late nineteenth and early twentieth centuries. In 1897, in Toronto, John Scott Keltie even chided his hosts for their lack of progress in geographical matters in comparison with what was done elsewhere, notably in regard to Britain's Arctic endeavours:

> But has not Canada reached a stage when she is in a position to follow the maternal example still further? What has contributed to render the name of Great Britain illustrious than those great enterprises which for centuries she has sent out from her own shores, not a few of them solely in the interests of science? Such enterprises elevate a nation and form its glory and its pride. Surely Canada has ambitions beyond mere material prosperity, and what better beginning could be made than the equipment of an expedition for the exploration of the seas that lie between her and the Pole? I venture to throw out these suggestions for the consideration of those who have at heart the honour and glory of the great Canadian Dominion.[55]

Recognising in 1909 that 'good maps and geographical education are of use to all countries', Colonel Sir Duncan Johnston urged Canadians to map themselves in order better to know themselves. Such standard mapping, he stressed, was essential to 'bind together' such a thing as the British Empire 'whose different parts are geographically so scattered, but which are so closely bound together by common ties of kinship, interest, sentiment, and loyalty'.[56]

The exception to this trend was the address by Sir Charles Lucas, given in Adelaide on 12 August 1914. Lucas, who in his *Greater Rome and Greater Britain* (1912) had likened imperial Britain to imperial Rome, and who wrote several historical geographies of the British colonies, talked on 'Man as a Geographical Agency'. His attention to man as a geographical subject in contrast to the more common discussion of geography's physical content is noteworthy. So, too, was his interest in the geographical 'civilisation' of Australia through irrigating the desert and reducing that country's relative isolation: 'if you come to think of it, what geography has been more concerned with than anything else, directly or indirectly, is distance'.[57] As one newspaper report had it in advertising Lucas's speech – in tones which speak to that shift in geography's exploratory cultures noted in Chapter 3 – new environments needed new geographies:

Geographies of empire 121

> The argument which he will develop is that the age of geographical discovery is passing away, and that geography will in future be largely concerned with the changes which men – notably scientific men – have made and are likely to make. . . . He will try to point out how, over and above recording actual physical changes, science is in effect making them, that the quickening of speed in effect reduces distance, so that miles do not mean the same thing, and that islands will not be islands in the present sense, if flying becomes a regular mode of communication, in fact, that science, directly or indirectly, has profoundly altered geography, and is likely to alter it still further.[58]

As many delegates came to appreciate, knowing about imperial geographies depended upon closing the distance between themselves and others – geographically and epistemologically – by seeing things for themselves.

Encountering empire: excursions

Each of the seven overseas meetings involved major excursions in transcontinental rail travel after the paper sessions. In the South African and Australia meetings, the BAAS visit as a whole was structured around mobility: moving delegates from one town to another allowed Dominion institutions to present themselves as credible authorities for the conduct of science, and allowed locals to present sites of scientific importance and ensure that BAAS visitors could establish their own scientific status as they encountered the geographies of empire for themselves (Figure 4.3). BAAS excursions thus further illustrate the centrality of empirical encounter to the local making and conduct of science. Equally, they highlight the limitations to scientific knowledge inherent in fieldwork: mobility provides immediacy but no time for measured reflection; encountering the novel, memory is no secure guide to perception; individual 'type' sites provide no basis for generalisation; fieldwork (time-limited) depends upon the supposed immutability of accumulated specimens and upon the presumption of a single and shared language of standardisation and culture of experimental replication.[59]

Mobile exploratory science was even a feature of the overseas meetings in the making, as it were, as BAAS scientists and visitors travelled overseas. For A. J. Herbertson, heading for South Africa in 1905, shipboard life allowed everyone to be a scientist:

> Never has there been a more impressive demonstration of the advance of the material sciences! To its success workers in every section have contributed, from the astronomers and magneticians of A to the schoolmasters of L. If the engineers have combined the results of many investigations in constructing the ways and the means of transportation, geographers, often disguised as sailors and engineers, have discovered and investigated

Figure 4.3 Excursion science at Niagara Falls, 1884

the paths followed. Every traveller becomes for the nonce a geographer, and we may regard the work of the geographical section as lasting for ten or twelve weeks, and all the 'over-seas' party to have been members of it. Instead of calling the expedition the greatest picnic on record [*cf.* Simpson], it might with greater truth be described as the most remarkable geographical excursion ever carried out.[60]

The leading American geomorphologist William Morris Davis spent a total of fifty-eight days at sea in attending the 1905 meeting. He helped organise lectures from amongst the travelling delegates: 'hardly a day passed on shipboard without a group of devotees learning new tricks'.[61] For others on board the *Parisian* to the 1897 Toronto meeting:

> Much amusement was caused by games of various kinds; the anthropologists bringing out their instruments and measuring the directions of skulls in every direction – mine, by the by, being a very small one, & not capable of holding much! By way of instruction, we were all much interested in hearing Mr Selous the distinguished explorer, relate some of his African experiences which he did with much effect. The greatest excitement however was caused by the first view of icebergs as we neared the coast of Labrador.[62] [Figure 4.4]

Such science and geography on the move allowed BAAS delegates to see the colony in question, pursue specialist scientific work within

Geographies of empire

Figure 4.4 BAAS delegates en route for the BAAS meeting, Toronto, 1924, on the SS *Caronia*

limits and be viewed by and appealed to by residents. In the Canadian meetings especially, the importance of seeing colonial development at first hand was stressed repeatedly: 'To the geographers especially the opportunity of seeing a country in the making, and at the same time of coming in to contact with those who are moulding its destinies and developing its resources, was an unforgettable experience.'[63] Managing that experience was important. As a local newspaper observed of the Toronto excursions in 1897:

> The visitors did not come in a body, but in scattered groups. You cannot decorate a British scientist with a badge, whip him into a line of basket-laden excursionists, and make him admire things as they are pointed out to him; he wants to see things for himself, and the local authorities, by allowing him to do so, while assisting him to view the sights in the most pleasant way, showed good judgement.[64]

As they encountered such new geographies, so visiting geographers emphasised the importance of their subject in colonial development: O. H. T. Rishbeth, from the University of Southampton, for example, talked to a business gathering in New Westminster, British Columbia, explaining:

> the nature and functions of modern geography as a regime of scientific thought and investigation, and then dwelt . . . upon its functions and potentialities for service in young communities in rapidly developing

territories. Special emphasis was laid on that practical aspect of the subject under which geography may be regarded as the science of settlement.[65]

Excursions afforded directive instruction from colonial scientists or others to BAAS visitors. Visiting experts were, fleetingly, novitiates in unknown geographical settings even as they were also expected to take home new findings in order to benefit colonial science and assist in the politics of immigration and economic development. In Australia in 1914, organisers were keen to use the visit to 'prosecute Biological and Anthropological Research in Oceania by means of an expedition to northern Queensland including some of the representatives of those branches of Science attending the meeting': making such things work properly depended, as Rivett termed it, on excluding 'the mere pleasure seekers'.[66] But as J. W. Gregory of the University of Glasgow cautioned Rivett, 'it seems to me that the Australian committees cannot know the most suitable places in such subjects as engineering, geology, geography, botany, zoology and agriculture until they know who is coming. The excursions will no doubt be selected to suit the average, but some of the men who go out may want to see some special locality.'[67] Rivett even tried to manage excursions by topic and gender: 'Ladies cannot go on geological excursions: these may be tough, wet, and arduous. For all other excursions, sex is a matter of indifference.'[68] Local circumstances could militate against geographical understanding in other ways – section E delegates on an excursion near Cape Town in 1929 bewailed 'the want of expert assistance in the study of the extraordinarily interesting local flora'.[69]

One consistent theme of the overseas excursions – where geography often came together with section H (anthropology) and, sometimes, section I (physiology) – was 'study' of colonial native peoples. Whilst serious ethnographic enquiry was an established feature of the BAAS (see Chapter 5), overseas excursions could produce a sort of 'ethnological pageantry' of the indigenous peoples. In 1884, one Canadian ethnographer reported upon 'the reluctance of the Indians to submit themselves to the process of measurement, or even, under satisfactory conditions, to the camera'. Things had altered greatly by 1909: at a ten-minute stop in Gleichen, Alberta, 'arrangements had been made for representatives of a tribe of Indians from a neighbouring reservation meeting the train at that point, arrayed in the gayest native costume. Contrary to their usual practice, they consented in the freest manner to be snap-shotted by the members of the Association armed with cameras.'[70] By 1924, natives were in readiness at railway stations as BAAS delegates arrived (Figure 4.5).

Figure 4.5 BAAS delegates encounter the 'Other': North American Indian natives on the BAAS trans-Canada excursion, 1924

In South Africa, in 1905, discussions about 'the best place for holding a Kafir dance, or for observing native life and customs' (finally held near Henley Station in the Zwaartkop) noted also that 'the Committee should be prepared to go to some expense in extending hospitality to the Natives in the way of giving them, say, a few head of cattle to kill, bread and sugar and salt, say, for the women and girls': weddings between tribes people were timed to allow BAAS visitors to witness such customary practices.[71] For Gertrude Mabel Rose, who witnessed the event in which over a thousand natives danced – 'the scene was most barbaric, the natives dressed very variously ... most of them wearing the scantiest apology for a costume'[72] – the spectacle was later overshadowed by her sight of the Boer War battlefields.

Overseas excursions were thus carefully orchestrated events – stage-managed, even – depending on who was coming and the value of science to dominion economies. For geography, excursions as a route to science provided a particular space for settlement studies, regional description and the study of human–environment interactions in colonial context. For colonial figures, excursions provided a means to weaken the tyranny of scientific distance.

Conclusion

On 3 December 1929, at a joint meeting in London of the BAAS and the Royal Empire Society, Sir Richard Gregory, editor of *Nature* and a leading figure at the South African meeting that year, spoke on 'Science and the Empire'. In his conclusion, Gregory noted the stimulus given to science in South Africa by the BAAS visit and the interests created among the visitors in the country's scientific and economic problems:

> My general survey represents not only personal conclusions as to the place of science in the development of the Empire, but also the views of competent authorities upon scientific authorities in South Africa. We all aim to advance progress and promote human welfare, and many of us are convinced that this can be attained only by scientific guidance. Upon statesmen and administrators is the responsibility of seeing that this guidance is rightly regarded and effectively used.[73]

For Gregory as for others before him science was part of a new commonwealth or dominion, intellectually, politically and geographically: as Sir William Bragg had noted of the 1924 Toronto meeting, 'There is a great similarity between the work of the scientists and the exploration of a new country. Both speak to the onward march of civilisation.'[74]

This chapter has outlined how science in the form of the BAAS and in one particular form, geography, worked to serve this imperial agenda between 1884 and 1929. But for other circumstances, BAAS overseas meetings would have continued after 1929: plans for a 1934 New Zealand meeting were put forward and a 1938 Australia meeting was also considered.[75] To those involved, as local organisers or visiting delegates, BAAS overseas meetings worked to advance science and to extend geographical knowledge in particular places and certain countries and to do so above national interests. In this last sense, some were more successful than others: the 1914 Australia meeting was curtailed (and planned excursions to Tasmania and New Zealand cancelled) by the outbreak of European war:

> Just at the moment when we were congratulating ourselves that culture and the study of the Arts and Sciences were breaking down the barriers of distance and almost of nationality – just when we were receiving in our midst as our guests the delegates of the British Association for the Advancement of Science, and several prominent German, French, and other foreign scientists, hell seemed to break loose.[76] [Figure 4.6]

The international brotherhood of science here fractured under the impact of European conflict.

In considering how the BAAS was bound up with imperial concerns, this chapter has demonstrated not just the close connections between geography, science and knowledge about empire. It has also illustrated the settings and the practices in and through which geography was put to work as a science of empire. In earlier meetings, geographical work on the colonies was done more by visiting experts. By the later meetings, colonial figures dominated. Section E proceedings – and those of the other sciences – were managed to ensure a focus on colonial topics even as, at the same time, those topics were used to advertise the colonies as spaces of economic and scientific opportunity.

If this is to offer correctives to the accounts of Pancaldi, Worboys and Dubow, it is also to highlight more general issues regarding the historical geography of geography as a form of science and at the transnational and local scales. Rather than simply chart the productive and receptive settings of science and geography, we may also consider the means by which they were made, travelled and engaged with by others. This may be a matter of scale and of practice more than site, of connections between social spaces and scientific institutions over long distances as well as of procedures within them. Questions of geographical scale matter in interpretation of how the BAAS worked since questions of how civic science worked overseas are inseparable from where and in whose interests. One consequence, which lends strength to the evidence above

Figure 4.6 'New Zealand's Call to the Man of Science.' The cover of a promotional leaflet issued by the New Zealand authorities to the BAAS. The call never evoked a response from the BAAS: plans to visit New Zealand as part of the 1914 Australia meeting were curtailed by the outbreak of World War I, and plans of 1934 were hindered by lack of finance

of the British experience, is that the BAAS itself ought not to be seen as a single body. Different sections, practitioners and epistemic cultures could and did give shape to what the science of empire was. Another and related consequence of examining the overseas meetings is that neither the nation, nor the city nor the town as a simple urban setting is alone a sufficient unit of analysis in the historical geography of imperial science. Networks of imperial exchange have been here highlighted: between cities and within subject sections, between people entrusted with demonstrating science in overseas settings and taking home news of colonial capacities and opportunities. Later overseas BAAS meetings built upon earlier experience: as *The Times* reported of the 1914 Australian visit:

> This visit adds completeness to that system of exchange of scientific thought and scientific men between the United Kingdom and the Dominions overseas which has been steadily growing since the beginning of the century; and it will serve to link still more closely the newer universities with the older throughout the whole of the Empire. The previous visits of the British Association to Canada and to South Africa gave fresh impetus to scientific work, and were followed by tangible results in the creation of new scientific institutions and laboratories. There is assurance beforehand that this precedent will be repeated in Australia, for one of the objects of the visit is to afford opportunities for joint discussion on cooperation in research between the southern continent and the countries of the Northern Hemisphere in astronomical, meteorological, and anthropological problems.[77]

This transnational geography of nascent colonial expertise was not only nurtured into maturity by British scientists. Science was also used to shape empire as, in turn, the colonial settings encountered by the BAAS helped shape what science, including geography, then was and would become.

Notes

1 J. Y. Simpson, 'The South African meeting of the British Association', *Scottish Geographical Magazine* 20 (1905), pp. 637–52, quote from p. 637.
2 This appears in the objects of the BAAS at its foundation, in R. Macleod and P. Collins (eds), *The Parliament of Science: The British Association for the Advancement of Science, 1831–1981* (Northwood: Science Reviews, 1981), p. i.
3 See, for example, B. Latour, *Science in Action: How to Follow Scientists and Engineers through Society* (Milton Keynes: Open University Press, 1987) and his notions of 'centres of calculation' and 'cycles of accumulation'. For examples and discussions of this model of internationalisation at one time or another in the British context, see J. Gascoigne, *Science in the Service of*

Empire: Joseph Banks, the British State and the Uses of Science in the Age of Revolution (Cambridge: Cambridge University Press, 1998) D. Miller, 'Joseph Banks, empire, and "centres of calculation" in late Hanoverian London', in D. P. Miller and P. H. Reill (eds), *Visions of Empire: Voyages, Botany, and Representations of Nature* (Cambridge: Cambridge University Press, 1996), pp. 21–37; R. Drayton, *Nature's Government: Science, Imperial Britain and the 'Improvement' of the World* (New Haven CT: Yale University Press, 2000); T. Richards, *The Imperial Archive: Knowledge and the Fantasy of Empire* (London: Verso, 2003); A. Stoler, 'Colonial archives and the arts of governance', *Archival Science* 2 (2002), pp. 87–109; R. Craggs, 'Situating the imperial archive: the Royal Empire Society library, 1868–1945', *Journal of Historical Geography* 34 (2008), pp. 48–67; H. Jöns, 'Academic travel from Cambridge University and the formation of centres of knowledge, 1885–1954', *Journal of Historical Geography* 34 (2008), pp. 338–62. For a discussion of the Eurocentrism in such work, see K. Raj, *Relocating Modern Science: Circulation and the Construction of Knowledge in South Asia and Europe, 1650–1900* (Basingstoke: Palgrave, 2007).
4 For discussions of this idea of knowledge circulating without restriction in such a Republic of Letters, see L. Daston, 'The ideal and reality of the Republic of Letters in the Enlightenment', *Science in Context* 4 (1991), pp. 367–86; C. W. J. Withers, *Placing the Enlightenment: Thinking Geographically about the Age of Reason* (Chicago: University of Chicago Press, 2007).
5 On these points, see E. Crawford, T. Shinn and S. Sörlin (eds), *Denationalizing Science: The Contexts of International Scientific Practice* (Dordrecht: Kluwer, 1993); L. Pyenson, 'An end to national science: the meaning and the extension of local knowledge', *History of Science* 40 (2002), pp. 251–90. For an illustration of shared (and uncertain) international collaboration around a thematic scientific 'moment', see C. Collis and K. Dodds, 'Assault on the unknown: the historical and political geographies of the International Geophysical Year, 1957–1958', *Journal of Historical Geography* 34 (2008), pp. 555–73.
6 G. Pancaldi, 'Scientific internationalism and the British Association', in MacLeod and Collins (eds), *The Parliament of Science*, pp. 145–69; R. MacLeod, 'Retrospect: the British Association and its historians', in MacLeod and Collins (eds), *The Parliament of Science*, pp. 1–16.
7 Pancaldi, 'Scientific internationalism and the British Association', p. 163.
8 M. Worboys, 'The British Association and empire: science and social imperialism, 1880–1940', in MacLeod and Collins (eds), *The Parliament of Science*, pp. 170–87.
9 S. Dubow, 'A commonwealth of science: the British Association in South Africa, 1905 and 1929', in S. Dubow (ed.), *Science and Society in Southern Africa* (Manchester: Manchester University Press, 2000), pp. 66–99.
10 Worboys, 'The British Association and empire', pp. 173–4.

11 MacLeod, 'Retrospect', pp. 1–2.
12 [Clara Strutt, Lady Rayleigh], *The British Association's Visit to Montreal, 1884: Letters by Clara Strutt Lady Rayleigh* (London: printed privately, 1884), Introduction.
13 MS. Dep. BAAS 197, Annual Meeting, South Africa 1905.
14 [Clara Strutt, Lady Rayleigh], *The British Association's Visit to Montreal*, Letter 4, The Opening Meeting, 26 August 1884.
15 University of Toronto Archives, MS. B65-0021 (02).
16 E. McLaughlin-Jenkins, 'Common knowledge: science and the late Victorian working-class press', *History of Science* 39 (2001), p. 458.
17 As J. Y. Simpson put it, 'There is little doubt that one great benefit of the Association's visit to South Africa lies simply in the fact that a number of intelligent British men and women were enabled to see vexed matters on the spot and obtain information at first hand, which was bound profoundly to modify any preconceived ideas on such topics as the native question, the education question, and Chinese labour': J. Y. Simpson, 'The South African meeting of the British Association', p. 641.
18 MS. Dep. BAAS 420, Press cuttings, 1905–1908, *The Times*, 24 July 1905.
19 Cape Town Provincial Archives, 3/CT 4/4/1/31 HM 217/4, Reception for the Meeting of the BAAS, July 1929.
20 National Archives of Australia, CP103/11, 41, Prime Minister's Department. This point about the connections between Australian national identity and involvement in international science is endorsed in the essays in R. W. Home and S. G. Kohlstedt (eds), *International Science and National Scientific Identity: Australia between Britain and America* (Dordrecht and London: Kluwer, 1991).
21 City of Sydney Archives, File 1910-1256, 24 June 1910.
22 MS. Dep. BAAS 423, Press cuttings, 1914–1916, 1919, *The Age*, 22 April 1910.
23 Adolph Basser Library, Australian Academy of Sciences A. D. C. Rivett papers, MS. 83, Correspondence Files MS83/29 3-3, Herdman to Rivett, 16 August 1913.
24 University of Toronto Archives, MS. B65-0021 (02).
25 R. L. Richardson, *Report of the Visit of the British Association to the Canadian North-West: Description of the Trips to the Rocky Mountains, Addresses Presented, Report on Speeches Delivered, Doings in Winnipeg* (Winnipeg: no publisher, 1884), p. 3. On the connections between geography, empire and Canada as an emigration field in this period, see A. M. C. Maddrell, 'Empire, emigration and school geography: changing discourses of imperial citizenship, 1880–1925', *Journal of Historical Geography* 22 (1996), pp. 373–87.
26 D. Knight, 'Tyrannies of distance in British science,' in Home and Kohlstedt (eds), *International Science and National Scientific Identity*, pp. 39–55.
27 S. E. Dawson (ed.), *Handbook for the Dominion of Canada* (Montreal: Dawson Brothers, 1884), p. 1.

28 Publication Committee, *Handbook of Canada* (Toronto: Publication Committee of the Local Executive, 1897), p. ix.
29 C. A. Lane, *A Guide to the Transvaal* (Johannesburg: Bartholomew & Lawler, 1905), p. vi.
30 G. Taylor, 'The physical and general geography of Australia', in G. H. Knibbs (ed.), *Federal Handbook, prepared in Connection with the Eighty-fourth Meeting of the British Association for the Advancement of Science, held in Australia, August, 1914* (Melbourne: Federal Government, 1914), pp. 86–121.
31 W. Flint and J. D. F. Gilchrist (eds), *Science in South Africa: A Handbook and Review* (Cape Town, Pretoria and Bulawayo: T. Maskew Miller, 1905), p. 1.
32 This is borne out in comments from the Office of the Governor General in 1929 comparing the state of affairs in 1929 with 1905: National Archives Repository, Pretoria, MS. GG 208 3/4902.
33 Local Publications Committee, *Handbook and Guide to Western Australia* (Perth WA: W. M. Simpson, 1914), p. 1.
34 See the Minute Books, Royal Geographical Society of Australia, Queensland Branch, Vol. 5 (1907–1916), pp. 199–216. The correspondence between W. A. Herdman and A. D. C. Rivett contains several references to the need for standardisation being thwarted by authors' views of the importance of their work on, for example, economic botany, geology or oceanography relative to the overall scientific depiction of the country: Adolph Basser Library, MS. 83/29 1–3, *passim*. On the Queensland body, see R. A. Butlin, '"Geography: imperial and local": the work of the Royal Geographical Society of Australasia (Queensland Branch), 1885–1945', in R. W. Childs and B. J. Hudson (eds), *Queensland: Geographical Perspectives* (Brisbane: Royal Geographical Society of Queensland, 2006), pp. 217–42. I acknowledge with thanks the help of Professor Robin Butlin for this reference.
35 R. A. Butlin, 'Historical geographies of the British empire, *c.* 1887–1925', in M. Bell, R. A. Butlin and M. J. Heffernan (eds), *Geography and Imperialism*, pp. 151–88.
36 MS. Dep. BAAS 202, fols 167–76.
37 RGS, Correspondence Block CB 1881–1910, A. J. Herbertson, 1 September 1905.
38 MS. Dep. BAAS 420, Press cuttings, 1905–1908, *Morning Post*, 30 October 1905.
39 [Anon.], 'Geography at the British Association', *Geographical Journal* 10 (1897), pp. 471–6.
40 MS. Dep. BAAS 326, Section E Minute Books, 1910–1920, fol. 9, 8 February 1909.
41 MS. Dep. BAAS 421; *Free Press News Bulletin*, 21 August 1909.
42 MS. Dep. BAAS 421; *Manitoba Free Press*, 2 September 1909; *Calgary Daily Herald*, 8 October 1909, p. 493.
43 R. Rudmose Brown, 'The British Association in Australia', *Scottish Geographical Magazine* 30 (1914), pp. 631–5, quote at p. 631. This

claim is echoed in J. McFarlane, 'Geography at the British Association', *Geographical Journal* 45 (1915), pp. 147–51: 'Among the papers given to the geographical section the larger proportion were contributed, as was appropriate, by Australian geographers and scientists, and the meeting has substantially added to intelligent appreciation of the geographical features and the potentialities of Australia' (p. 151).
44 Adolph Basser Library, MS. 83/26 3–6, Masson to Rivett, 27 November 1913.
45 MS. Dep. BAAS 236, Annual meeting, Australia: Correspondence, 1913–1914, Howarth to Herdman, 19 July 1914.
46 Adolph Basser Library, MS. 83/26 3–7, Rivett to Masson, 28 November 1913.
47 This point is stressed by one 1929 local committee, for example: Western Cape Provincial Archives and Records Services, MS. 3/CT 4/4/1/31 HM 217/4, Minutes of the BAAS Executive Committee (Cape Town Section), 20 June 1929.
48 Wits University Manuscript Collections, South African Association for the Advancement of Science, MS. AF 1211, Ad (1929); *Natal Mercury*, 21 August 1929. The connections between Natal's mineral wealth and visiting scientific expertise had been stressed for the first BAAS visit to South Africa: Robert Ababrelton, Secretary to the Natal Local Committee in 1905, noted how 'the mineral resources of Natal are practically unknown to the scientific men of the present day, and if they become interested in the mineral wealth of Natal and the development of the Colony in other respects, the flow of capital into the Colony will be increased, and its future prosperity be more assured': Pietermaritzburg Archives Repository, CSO 1780 1905/469, Ababrelton to Bird, 17 January 1905.
49 Of the twenty-eight Section E papers presented at the 1929 meeting, nineteen were concerned directly with South African geography: amongst the remainder, some – such as Marion Newbigin's paper 'The Mediterranean Climatic Type' – made comparative reference to it: MS Dep. BAAS 259. In 1905, only eight of the twenty-one papers presented to section E were on South African topics, although several papers of the 'greatest geographical interest' were read to other sections: A. J. Herbertson, 'The visit of the British Association to South Africa', *Geographical Journal* 26 (1905), pp. 632–41, quote at p. 641.
50 Withers, Finnegan and Higgitt, 'Geography's other histories?', pp. 438–40.
51 MS. Dep. BAAS 326, Section E Minute Books, 1910–1920, fols 58–9, 18 August 1914.
52 MS. Dep. 331, MS. Dep. 333. The 1933 report of the Committee on the Dominion Universities, presented at the Leicester meeting that year, outlines the poor state of geography in the universities of the Dominion (see also Chapter 7).
53 MS. Dep. BAAS 188–93.
54 MS. Dep. BAAS 326, Minute Books of Section E, 1910–1920, fol. 88, 11 September 1919.
55 J. S. Keltie, 'Section E: Geography', *Report of the Sixty-seventh Meeting*

of the British Association for the Advancement of Science (London: John Murray, 1898), pp. 699–712, quote at pages 708–9.
56 D. Johnston, 'Section E: Geography', *Report of the Seventy-Ninth Meeting of the British Association for the Advancement of Science* (London: John Murray, 1909), pp. 517–28, quote at p. 528.
57 C. P. Lucas, 'Section E: Geography', *Report of the Eighty-fourth Meeting of the British Association for the Advancement of Science* (London: John Murray, 1915), pp. 426–39, quote at p. 438.
58 MS. Dep. BAAS 423, Press cuttings, 1914–1916, 1919, *Morning Post*, 27 March 1914.
59 On these points, see for example J. Golinski, *Making Natural Knowledge: Constructivism and the History of Science* (Cambridge: Cambridge University Press, 1998), and the essays in M.-N. Bourguet, C. Licoppe and H. O. Sibum (eds), *Instruments, Travel and Science: Itineraries of Precision from the Seventeenth to the Twentieth Century* (London: Routledge, 2003).
60 Herbertson, 'The visit of the British Association to South Africa', p. 633. On the ship as a scientific site/instrument, see R. Sorrenson, 'The ship as a scientific instrument in the eighteenth century', *Osiris* 11 (1996), pp. 221–36.
61 Archives of the Association of American Geographers, Part I, Box 156 [W. M. Davis] 'A Visit to South Africa with the British Association, 1905'.
62 MS. Dep. BAAS 186, fol. 62.
63 [Anon], 'The British Association meeting in Canada', *Scottish Geographical Magazine* 40 (1924), pp. 345–56, quote at p. 345.
64 MS. Dep. BAAS 418, *The Globe* 23 August 1897.
65 [BAAS], *Narrative of Official Journeys in Canada in Connection with the Toronto Meeting, 1924*, London, 1924, 485.
66 Adolph Basser Library, MS. 83/30 1–6, 2 May 1913; 3–3 14 November 1913.
67 Adolph Basser Library, MS. 83/30, 1–6, J. W. Gregory to A. D. C. Rivett, 18 September 1913.
68 MS. Dep. BAAS 234, fol.205.
69 [M. Newbigin], 'The South African meeting of the British Association', *Scottish Geographical Magazine* 45 (1929), pp. 337–49, quote at p. 343.
70 MS. Dep. BAAS 190; [anon.], 'Geography at the British Association', *Scottish Geographical Magazine* 25 (1909), pp. 577–85, quote at p. 584.
71 Pietermaritzburg Archives Repository, SNA I/1.323 1905/1751, C. W. P. Douglas de Fenzi to S. O. Samuelson, 18 May 1905.
72 Wits University Manuscript Collections, Diary of Gertrude Mabel Rose, MS. A.1753, 60 [25 August 1905].
73 MS. Dep. BAAS 259.
74 MS. Dep. BAAS 426; *Montreal Gazette*, 22 August 1924.
75 MS. Dep. BAAS 282, Correspondence concerning Invitations to hold Annual Meetings, 1929–1952, *passim*, especially fols 6, 183.
76 W. J. Dakin, [Presidential Address, June 1915], *Journal and Proceedings of the Royal Society of Western Australia* 1 (1915), p. 227.
77 MS. Dep. BAAS 423, *The Times*, 29 May 1914.

5

Hierarchy, distribution, connection: geography as a science of the physical world

Having shown how BAAS meetings worked at home and abroad and addressed geography's content, my focus in what follows is with geography's place as a science, and, in particular, with the features that distinguished geography as a science of the physical world in the eyes of the BAAS and others. In developing earlier arguments about geography's sectional history and cognitive content, this chapter is an account of geography's relative hierarchical status rather than with its contents alone. Geography's place as a physical science was both achieved and ascribed: in the first sense, by the topics it addressed and how, in the second in its position within the institution but also, given the development of new specialist and disciplinary forms of knowledge from the 1830s on, within science as a whole, not just within the BAAS. Since scientific identity was attained, made in relation to what practitioners within certain subject areas looked at and how they looked, what came to be geography the science was defined by its practices and methods as well as by its relative institutional position. In exploring geography's epistemic distinctiveness as well as its hierarchical position, this chapter looks only at the physical world: geography, the human sciences and human geography are the subject of the next chapter.

My contention, simply, is that in so far as it concerned study of the physical world geography within the BAAS came to be seen as a science of distribution, its methods observational, descriptive and exploratory, the map its principal discursive form of presentation. As I hope to suggest, accounts of geography's sectional status as a science and the practices by which it came to distinguish itself as a science, within and outwith the BAAS, are better understood as relational rather than absolute things. BAAS personnel and sections worked and managed funds together (as we have seen for section E). Questions of distribution as method and as representation via the map were not unique to geography (being shared particularly by the botanical and the geological sciences). Geographical questions, about the variation of phenomena over space, for example, were seen as foundational to the search for causal connections in other sciences as their practitioners studied, in various ways, the empire of nature. In short, geography came to be seen as a distributional practice – what is where? And why? – and as a foundation for other more causally focused and law-based sciences.

I begin by paying attention to geography's place in the hierarchy of the sciences in the BAAS. Two related themes are then examined to elucidate geography's development as an exploratory science and as a science of distribution. The first concerns the issue of terrestrial magnetism, what was known then as the 'Magnetic Crusade', and the later neglected history of global geodetic survey within the BAAS. In this particular respect, my aim is to extend what is known of the Association's work and institutional involvement in the topic and to highlight the emphases placed upon fieldwork and observation within science's exploratory cultures. The second is that of topographic mapping. Accurate measurement and delineation of space in the form of the map were seen as essential to the development of scientific knowledge as a whole and a defining practice for geography. The fact that the debates over terrestrial magnetism and the importance of mapping both took definitive shape from the mid 1830s is not coincidental, for it was then, and in the work of the leading gentlemen of science – for geography notably the figure of Sir Roderick Murchison – that geography and the sciences were being put to order.

Questions of hierarchy and of distribution: 'geography emergent'[1]

As Morrell and Thackray demonstrate for the BAAS before about 1845, questions of scientific identity, understood in terms either of what participants did or audiences thought or of one subject's place relative to another cannot easily be understood without attention to the politics and hierarchies of science within the BAAS, and to how the BAAS worked over time to promote and develop itself.

The mathematical sciences were widely accepted as pre-eminent, in the BAAS and more widely, because of the personnel involved and because of the concern to identify universal laws and to determine causal connections in the physical world. As Charles Daubeny put it in 1836, 'All the physical sciences aspire to become in time mathematical: the summit of their ambition, and the ultimate aims of the efforts of their votaries, is to obtain their recognition as worthy sisters of the noblest of these sciences – Physical Astronomy.'[2] But not all the sciences offered the identification of mathematical laws – or even mathematical treatment of their subject material – as a defining characteristic or end in view. Endersby has shown of botany, for example, how its essential characteristic, classification, meant that the subject was always low in the hierarchy of the BAAS sciences and, thus, of science as a whole, at least in Britain. This was evident in the very low levels of funding given to section D, natural history, which incorporated botany, between 1835 and 1856 (8 per cent of overall BAAS sectional expenditure) in comparison with that apportioned to the mathematical and physical sciences (38.5 per cent) in the same period. (Geography, like botany, only received 8 per cent in this period.) It was apparent in the numbers of published reports by section between 1835 and 1844, where section D produced twenty-six reports, 18 per cent of the overall total of 144 sectional reports, in contrast to the fifty-eight (40 per cent) from section A (and the seventeen reports (12 per cent) produced by what, in that period, was geography and geology, section C).[3] And while figures for sectional funding show that botany was always second best to zoology within natural history, this was also so because, in being concerned with classification, practitioners were not thought 'philosophical' enough in the eyes of contemporaries in terms of what Murchison in his 1846 BAAS presidential address considered 'the higher questions of structure, laws and distribution'.[4]

Yet, as Endersby also notes, distributional concerns provided scientific status of a sort for botany, since, if structure was the concern of zoology and the matter of law formulation never really within the reach of the botanical sciences, studies of distribution could mean looking at why plant assemblages featured in certain areas and, from that, could permit the search for potential causal connections between plant communities, climate, soil and altitude. This distributional emphasis within the botanical sciences and the necessity to connect plants with their environment thus brought botany and geography together (Figure 5.1). As the author of one regional flora noted:

> A large and increasing share of attention has of late been paid, by Botanists, to that branch of the science which has been denominated *Local botany*. ... The publication of these works, by diffusing more widely a knowledge of the local distribution of plants, has tended to the still farther promotion of that branch of Botany called Botanical Geography.[5]

Others have documented how such local floristic geographical work was then being undertaken on the Continent in the work of Alexander von Humboldt and Alphonse De Candolle: indeed, 'Humboldtian science' was in part predicated upon identifying causal associational connections between the distribution of plant communities and physical geographical factors.[6] For one reviewer in 1856 – who preferred the term 'Geographical Botany' in referring to De Candolle's work – 'The

Figure 5.1 Section D, Biology, at the BAAS meeting, Manchester, 1887

investigation of these facts, of the presumed origin and subsequent migrations of plants, and of the causes which influence the phenomena observed, constitute those sciences in which the labours of the botanist are connected with those of the geographer and the geologist.'[7]

We can see quite how closely the emergent sciences and their constituent practices were connected from early debates over geography and geology. Geography, as we have noted, was initially placed within the Geological and Geographical Sub-committee. Section C, geology and geography, was dominated by geologists and was at first chaired by Sir Roderick Murchison. By 1838, the overbearing geological focus of the section was of concern to Murchison, as was the perception that the Royal Geographical Society, after only a few years of its existence, was even then overshadowing the BAAS's geographical mission:

> In noticing the labours of the Section of Geology and Geography, we have to observe, with regret, that the latter science has not hitherto received at our meetings that amount of attention to which it is justly entitled. When we consider the advances which the science has recently made under the auspices of the Royal Geographical Society of London, we cannot but lament that the British Association did not, at an earlier period, request a report from some one of its members upon the present state of our

geographical knowledge, and upon those departments of it in which our researches might be most advantageously prosecuted.[8]

Years later, in addressing the by then geography and ethnology section which he had been influential in establishing (see the following chapter), Murchison looked back on how geography early on had been 'almost submerged by the numerous memoirs of my brethren of the rocks'.[9] Murchison's remarks in the 1830s are an appeal to the subject's existence in more than one sense – institutionally and for more than one formal body – in terms of its range and methods, and in regard to geography's position and content relative to other sciences.

Geography's initial inferior position *vis-à-vis* the mathematical and the earth sciences is not to say that geography was not a concern at the outset for the BAAS. As Mill pointed out in 1931, 'The philosophers of 1831 founded their Association in this spot [York] as a result of reaction against geographical limitations.' For them, 'The value of Geography was present in the minds of the founders, as it could not fail to be when one of them was Mr. Roderick Murchison, fresh from the inauguration in the previous year of the Royal Geographical Society.' As Stafford and others have documented, Murchison did indeed occupy an important place within the RGS and within the BAAS in shaping the relations and purpose of section E: as Howarth notes, 'It is impossible to overrate the personal power of Murchison in the Section from 1851 to 1870.'[10] But 'geographical' in this sense of location and the urban provincial sites in which the BAAS did and would act was not the only consideration given to the question of geography. As the proceedings of the 1831 York meeting noted, 'there are many important questions in philosophy, and some whole departments of science, the data of which are *geographically* distributed and require to be collected by local observations extended over a whole country; and this is true not only of those facts on which single sciences are founded, but of many which are of more enlarged application'.[11]

At foundation, then, geography was seen as a distinct subject but perhaps less clearly that than as a practice – of determining the distributional significance of phenomenal variation over space, at local and supra-local scales – whose value lay in assisting other sciences. In the sense of practical exploration and survey mapping as a basis to other forms of knowledge, geography was from the outset recognised as a key theme in the Association's consideration of what was embraced by the term 'science' and was so on the basis of its attention to the distribution and location of phenomena. This point is reinforced in noting other 'definitions' of geography such as that proposed by Mary Somerville, whose *Physical Geography* (1848) embraced 'a description of the earth, the sea and the air, with their inhabitants animal and vegetable, of the distribution of organized beings, and the causes of their distribution'.[12] There were structural and substantive connections with geology, given the subjects' role in section C and practitioners' attention to the earth's age, its physical features and the relative positioning in place and over space of different rock types and what such spatial expression meant for an understanding of 'deep time'. There were connections with what Endersby has noted for botany and what contemporaries observed of 'botanical geography' or 'geographical botany'. Such connections help explain why, for example, Joseph Hooker (who worked on what he called 'physico-geographical or geographico-botanical districts' in his 1844–47 *Flora Antarctica*) addressed the 1866 Nottingham meeting on the geographical distribution of insular floras, why he spoke to the title he did in presiding over section E in 1881 (see Table 3.4), and why Richard Owen should address that section in 1858 on 'The Geographical Distribution of Plants and Animals' as part of his presidential address and there assert how 'Botany, in short, at this phase becomes intimately connected to climatology: and the traveler, the meteorologist, and the naturalist reciprocally aid each other in the acquisition of a knowledge of fruitful general laws'.[13]

In turning in more detail to two illustrative themes in the promotion of geographical work concerned with mapping, distribution and location, what is revealed is not just particular instances in the shaping of geography and its connection with other sciences within the BAAS, but also something of how science both presented and conducted itself within Britain and how national scientific work demanded and depended upon international collaboration and competition over the 'resource' and the practices embodied in doing science. The example of terrestrial magnetism illustrates the connectedness of the sciences.[14] The example of topographic mapping illustrates the dependence of the physical sciences upon a sense of geography as the study and practice of distribution: as was apparent at the outset, the Association's representation to the government in the 1830s over the work of the Ordnance Survey, notably in Scotland, centred upon its view that 'the progress of various branches of science is seriously retarded by the want of an accurate Map of North Britain'.[15]

Terrestrial magnetism, the 'Magnetic Crusade' and the geographies of geodetic survey, 1834–c. 1945

Existing studies on the place of terrestrial magnetism in nineteenth-century British science focus upon the so-called 'Magnetic Crusade' and its effective cessation by the later 1840s.[16] Yet terrestrial magnetism and collaborative geodetic survey remained a concern of the BAAS from the mid nineteenth century and into the twentieth. That they did so points, I suggest, not just to developments in the sciences so much as to a continued concern within the BAAS over its 'global reach' as an institutional instrument of science. In the Association's later attention to a global network of geodetic observatories, as well as in plans for its establishment before about 1845, we can see one expression of that agenda of scientific internationalism which the BAAS evinced at foundation. Indeed, the story of the Association's involvement with terrestrial magnetism before and after 1845 cannot be read without addressing the connections between matters of international rivalry, international collaboration and national and individual self-interest. It is an account which depended upon securing connections, in and between different locations, between scientific benefit, institutional authority and geographical setting – in short, a narrative of local, national and international geographies of science which helped establish, especially between 1834 and 1845, the authority of the BAAS itself.

BAAS and international interest in terrestrial magnetism had deep roots. The need for accurate measurement of the earth's magnetic variation was recognised as a significant problem centuries before the Association's involvement. The early modern natural philosopher William Gilbert had tackled the question in his *De Magnete* (1600) and Edmond Halley made important advances on the *Paramore* voyages at the end of the seventeenth century.[17] From the 1730s, terrestrial magnetism was bound up with what was perhaps the greatest geodetic question of all, the 'Shape of the Earth' debate, itself a reflection of differences between Newtonian and Cartesian natural philosophical principles. For Newtonians, the earth was an oblate spheroid – pumpkin-shaped – flattened at the poles; for Cartesians, it was otherwise: the earth was a prolate spheroid – melon-shaped – flattened at the equator. The testing of these opposing views required systematic in-the-field observation and verification. Expeditions to Lapland and to what is today Ecuador to determine magnetic declination confirmed Newtonian principles. But in being based on relatively few measurements, Enlightenment expeditionary science left geodesy as an open field of enquiry – one increasingly international in nature and conduct – by the turn of the nineteenth century.[18]

It was this long-standing interest in the question and the issues of international collaboration and national scientific standing it demanded and provoked that attracted the BAAS to terrestrial magnetism from 1831. As Cawood and others have shown, leading members of the BAAS at foundation saw investigations into terrestrial magnetism not just as a means to improve navigation but as a means to search, via expeditionary science and empirical observation, for solutions to nature's laws. And for figures like Vernon Harcourt, John Phillips, William Whewell, the polar explorer James Clark Ross, John Herschel and, notably, Edward Sabine, the pursuit of magnetic variability was a means to secure their status and that of the newly established Association: 'On the level of organizations, the fortunes of the Magnetic Crusade were inextricably bound up with the early development of the British Association for the Advancement of Science and reveal something of the ideology of that institution in its attitude to the funding of scientific enterprise and to the participation of its members.'[19] These men worked to do this, moreover, in the face of parallel developments in France, where the leading centre of magnetic research was in the Paris Observatory, and in Germany, where research was concentrated, after 1834, in Göttingen in the Magnetische Verein (Magnetic Institute).

The details of these international exchanges around terrestrial magnetism, competing explanatory theories, British insistence upon the search for natural laws confirmed by systematic observation in the field and the individuals involved in promoting the topic – the so-called 'Magnetic Lobby' and its supporters, together the 'Magnetic Crusade' – have been covered elsewhere and need not be repeated. Here I want to chart the ways in which debates on terrestrial magnetism were shaped in different BAAS annual meetings, and to consider how the Association's engagement with the topic came to depend not just upon the establishment of a network of observatories and measuring stations across the globe, but also upon certain sorts of exploratory scientific endeavour that came to be seen as geographical work. Given what we have seen to be the importance attached to exploration in geography's profile as a science within the BAAS, there are important connections, largely overlooked hitherto, between the success of exploratory voyages in the study of terrestrial magnetism, the status of exploration within section E and the scope and nature of geography as a science.

Geographies of terrestrial magnetism, 1831–1845

Initial moves to promote the observation of magnetic variations, within England, were made at the BAAS foundation meeting in 1831, and the topic was returned to in the two meetings following. Advances were made at the 1834 Edinburgh meeting when it was agreed that the Association would co-operate with French magnetists, and in Dublin in 1835, where a committee on magnetism was established and a resolution passed to stress to the government 'the importance of sending an expedition into the Antarctic regions . . . with a view to determining precisely the place of the Southern Magnetic Pole or Poles, and the direction and intensity of the magnetic force in these regions'.[20] Despite these initiatives – and their promotion within Dublin was lent support by the location in that city of Humphrey Lloyd's magnetic observatory – they failed because their proponents did not secure the support of either the Royal Society or of the government. Only with the involvement of Edward Sabine, a leading figure within the Royal Society, and with his completion of the magnetic survey of the British Isles – presented at Bristol in 1836[21] – and with the support of Roderick Murchison and of John Herschel after 1838 did the promotion of terrestrial magnetism begin in earnest.

At the 1838 Newcastle meeting, further resolutions to promote the systematic observation of magnetic variance were successfully advanced. Under Herschel's and Sabine's guidance, they attracted the necessary government support in the following year. The recommendations included provision of magnetic observations 'in the British colonies in co-operation with and in extension of the continental system of observatories', the claim that 'Important places to make measurements are Canada, Ceylon, St. Helena, Van Diemen's Land, and Mauritius, or the Cape of Good Hope', reiteration of the importance of a naval expedition to the Antarctic 'to make observations in the south polar region', and an insistence that 'Observations [i.e. magnetic measurements] should be made by naval officers on the expedition.'[22] By 1840, a system of magnetic observatories had been established, variously supported by the Admiralty and the Royal Artillery (at Greenwich, Dublin, Toronto, St Helena, the Cape of Good Hope and in Van Diemen's Land), the East India Company (at Madras, Simla, Singapore, Bombay), and in association with foreign governments and overseas scientific bodies (ten stations in Arctic and European Russia and thirteen other stations worldwide). The 1839–43 Antarctic expedition of HMS *Erebus* and HMS *Terror* under James Clark Ross made important contributions to studies of terrestrial magnetism (and to the zoology of oceans), studies which were developed by Sabine in later years. With the publication of his two-volume *A Voyage of Discovery and Research to Southern and Antarctic Regions* (1847), gold medals from the Royal Geographical Society and the Société de Géographie in Paris, fellowship of the Royal Society and a knighthood, Ross secured his public and scientific reputation, realised the vision of Sabine, Herschel and the other magnetists and, effectively, ended the 'Magnetic Crusade'. Interestingly, given this point about naval officers' in-the-field scientific credibility, the French and the Americans saw his work differently as based on inaccurate

Victoria Barrier and Land. Lat. 78° S. Mount Erebus (active Volcano), and Mount Terror.

Figure 5.2 The hazards of Antarctic science, as illustrated on the title page of Joseph Hooker's botanical work undertaken whilst on the expedition led by James Clark Ross 'for the purpose of investigating the phænomenon of Terrestrial Magnetism in various remote countries, and for prosecuting Maritime Geographical Discovery in the high southern latitudes'

observations made by untrained naval officers not trained scientists and more dependent upon shared endeavour than Ross admitted.[23]

Rather than see this period of scientific resolutions, political appeals, James Clark Ross's work and the establishment of the geophysical stations as an inevitable apotheosis of the magnetic lobby, other interpretations of its significance may be offered. Magnetic observatories and the study of terrestrial magnetism represented the global reach of British geodetic science – and its dependence upon in-the-field exploratory work and international collaboration. The hazards of polar science are hinted at in the illustration which forms part of the title page to Hooker's Antarctic work on the Ross expedition (Figure 5.2). Naval men provided the raw material for British expeditionary science, the Admiralty under Sir John Barrow the key institution in terms of funding science, especially in polar regions.[24] This first point was reinforced by Herschel in his editorship of the *Admiralty Manual of Scientific Enquiry*

(1849) where he noted that instructions over collection and procedure should be 'generally plain, so that men merely of good intelligence and fair acquirement may be able to act upon them'.[25] The nature of the science being funded and directed was based upon systematic measurement, repeated observations and the co-ordination of others' results – in short, a form of 'Humboldtian science'.[26] The Association, moreover, acted to co-ordinate the interests of those men with the authority to make sense of the results and who 'spoke' for magnetic science as, in turn, the newly founded body promoted itself. Public attention had already been swung towards polar work by the 1830s from newspaper reports of the HMS *Hecla* voyages of William Edward Parry between 1819 and 1825, from John Ross's search for the North West Passage between 1829 and 1833 and in the published narratives that followed in their wake. The reportage of such work in tones of dispassionate neutrality – polar science and its narration as cold fact – likewise helped establish both the naval explorer-cum-scientist and the account of his work as credible sources of new natural knowledge.[27] What is also clear is that in its emphasis upon exploratory science, its support of fieldwork and systematic attention to observation in the field it was the BAAS and the Admiralty, not the RGS, that provided from the early 1830s to at least the mid 1840s the epistemological 'definition' and the institutional base for geography as a science of exploration.

We have seen that the sectional status of geography within the BAAS was not clear at foundation. But geography's place as a science was recognised as part of the appeals of the magnetic lobby. This is clear from the proposals in Dublin in 1835. These noted:

> That a representation be made to the Government of the importance of sending an expedition into the Antarctic regions, for the purpose of making observations and discoveries in various branches of Science, as Geography, Hydrography, Natural History, and especially magnetism, with a view to determining precisely the place of the Southern Magnetic Pole or Poles and the direction and intensity of the magnetic force in those regions.[28]

The fact that such an expedition was not undertaken for a further four years does not diminish the claims made for geography as a science. Neither, too, should this statement about geography as a branch of science be over-read – as indicative of an established 'disciplinary' identity and epistemic core – not least given the uncertain and contingent nature of 'science' as a whole and what we have seen about geography's principal research themes.

Nevertheless, the sense that geography was recognised as part of science and that geographical knowledge about the earth was seen as contributing to scientific advance is clearly borne out by the remarks of Murchison and Sabine in Glasgow in 1840 (even as they there also acknowledged the prior magnetic and polar work of the French under Jules Dumont d'Urville and the Americans under Charles Wilkes):[29]

> Although terrestrial magnetism stood forward as the prominent object of the Antarctic expedition, yet it was also destined to advance our knowledge of the 'physique du globe', in all its branches, and especially in that of geography. Had the project of an Antarctic expedition been acceded to when it was first proposed, viz. at the meeting of the British Association, in Dublin, in 1835, there can be no reasonable doubt, that a discovery of coast, which by its extent may almost be designated as that of a Southern Continent, situated in the very region to which its efforts were to have been chiefly directed, must have fallen to its lot; and the flag of England [sic] been once more the first to wave over an unknown land. But while, as Britons, we mourn the loss of a prize which it well became Britain and British seamen to have made their own, it is our part too as Britons, as well as men of science, to hail the great discovery – one of the very few great geographical discoveries which remained unmade – and to congratulate those by whom it has been achieved, those whom we are proud to acknowledge as fellow-labourers, and who have proved themselves in this instance our successful rivals in an honourable and generous emulation.[30]

The centrality of geography is apparent, too, in Herschel's plea for the co-ordinated global geography of observation stations when, in Cambridge in 1845, he stressed how 'the gigantic problems of meteorology, magnetism and oceanic movements can only be resolved by a far more extensive geographical distribution of observing stations' and how each 'civilised nation' should consider its observatories as 'a local centre of reference for geographical determinations and trigonometrical and magnetic surveys'.[31] And it is evident in Murchison's presidential address the following year:

> Thus it was, when contemplating the vast accession to pure science as well as to useful maritime knowledge, to be gained by the exploration of the South Polar regions, that we gave the first impulse to the project of the great Antarctic expedition, which, supported by the influence of the Royal Society and its noble President, obtained the full assent of the Government, and led to results which, through the merits of Sir James Clark Ross and his companions, have shed a bright lustre on our country, by copious additions to geography and natural history, and by affording numerous data for the development of the laws that regulate the magnetism of the earth.[32]

There is, of course, much else that might be said of this evidence: BAAS involvement in national and international collaboration; its promotion of heroic polar endeavour and of magnetic work in general in order to establish its own institutional authority; the rhetorical claims of leading men of science, keen to see Britain take a leading role in helping solve a matter of global importance. Yet this is also an important appeal to geography: as an emergent subject, as a developing science, and to the role of geographical exploration and observation – to being and doing in the field – as a contribution to the advancement of science. It is quite in keeping with other evidence, Murchison's views in particular, about the importance of exploration and of mapping in geology as well as in geography, and about the place of the field-based sciences, notably geography, as agents of empire.[33] Consider, after all, his several presidential addresses to this effect (Table 3.4).

Studies in decline: terrestrial magnetism and geodetic science, c. 1845– c. 1945

Extant accounts of BAAS involvement with terrestrial magnetism rightly consider 1845 to be the year of 'peak' activity, with the network of geomagnetic observatories continuing for a further three years and Sabine publishing the results until the early 1850s.[34] But the story does not end there: the voyage of HMS *Challenger* between 1872 and 1876, for example, included magnetic measurements as part of its work. Terrestrial magnetism was part of the development of what became the oceanographical sciences in the nineteenth century, within the BAAS and more generally.[35] During the second half of the nineteenth century, however, the subject of terrestrial magnetism was increasingly subsumed within the work of section A, mathematics and physics, or pursued through the Royal Society. Further, the very things that had made it successful in the period 1834 to 1845 – polar expeditions, governmental support and institutional collaboration – were less easy to effect in later decades. This was partly because of the international co-ordination needed to secure consistent measurements from scattered observatories. And it was partly because of what doing the science required.

At the Dublin meeting in 1857, as part of a paper about the search for Sir John Franklin's lost expedition, Clements Markham spoke forcefully about the importance of polar work to geography and to science and of British naval men to the doing of such polar work: 'The Arctic Regions have ever been a field for the display of all the most heroic qualities of British seamen . . . the best and bravest worthies of our Navy have ever been ready to dare the hardships of those frozen regions, in pursuit of the noble objects connected with geographical discovery, and the advancement of science'[36] (*cf.* Figure 5.2). His plea was lent support by a resolution drawn up at Leeds the following year from Humphrey Lloyd on behalf of the BAAS General Committee to Lord Palmerston, the

Prime Minister, calling for renewed magnetic research in the vicinity of the Mackenzie river in the Canadian north-west. The resolution began by noting, 'Among the most important results of scientific research during the last twenty years, is the addition which has been made to our knowledge of Terrestrial Magnetism' and continued 'But, as always happens in real progress, the advance which has been made has shown new ground which ought to be explored.' Naval ships were available, the science was compelling – 'Terrestrial Magnetism had proved to be of high importance to science' – and the risks known but still worth taking – 'an expedition to this locality will be attended with no unusual risk'.[37] No answer was received.

This 1858 resolution and Markham's 1857 testimony to science's and geography's exploratory tradition, to the Navy's capacity and to hardship endured by human instruments at the margins of the earth has something of the rhetorical tone of Murchison and Sabine decades before. Markham was appealing to a wider interest, given the British public's engagement with polar exploration in the years between 1818 and 1860. But expeditionary science was expensive, polar work especially so. And although Markham did not know of it as he spoke (nor did anyone), the disaster that had by then befallen Franklin and which prompted public interest in the man and in the rigours of polar science – as we saw for the 1860 Oxford meeting – allowed the institutions involved to reappraise the cost. Government support was vital but could not be guaranteed. A joint resolution of 1873–74 on the need for an Arctic expedition from the RGS, the BAAS, the Royal Society and the Dundee Chamber of Commerce (which last stood to gain from the involvement of local crews) was initially rejected by Gladstone, but with the change of government the societies' principal officers approached Disraeli (only to be again thwarted). Writing to H. W. Bates of the RGS in May 1887, Admiral Sir Erasmus Ommanney (who, as the captain of HMS *Assistance*, had searched for Franklin in 1851) expressed his disquiet at the continuing difficulties facing expeditionary science:

> I had a conference with Sir J. Hooker, he could see no prospect of assembling the B.A. Antarctic Committee, – without the co-operation of the Royal Society and the R.G.S. and powerful parliamentary influence he sees no way to get a hearing for a Govt. Expedition, and there is no one at the Admrty Board who feels any interest in such matters.[38]

Even within section E, continued Ommanney, there was considerable indifference to the matter, and he later resigned himself to the likelihood that the RGS would take the lead in promoting such work.[39]

By the turn of the century, what decades before had been a crusade was by then marginal to the concerns of the BAAS, or at least to geographers. This was so much so that, in his presidential address to section E in 1903, Captain Ettrick Creak felt compelled to remind his audience of the importance of terrestrial magnetism to geography and of both to the vision of the BAAS. His talk was part appeal to the importance of terrestrial magnetism and to the memory of British seamen-scientists (of whom he was one) – citing Ross's work and the *Challenger* voyage – and part recognition of the limitations of the work accomplished and a plea for continued developments. Speaking of 'the elaborate surveys to date' – British work, the geomagnetic observatories and others' endeavours – Creak emphasised 'how much the results depend upon the nature of the locality' and how important it was to have consistency of measurement at the principal observing stations in order that worldwide sets of readings might be obtained and compared. His endorsement of the science in his closing remarks would have pleased Sabine, Murchison, Herschel and others:

> I also entertain the hope that geographers will become more interested in a subject so important to pure science and in its practical applications, and that it will become an additional subject to the instructions which travelers can now obtain under the auspices of the Royal Geographical Society in geology, botany, zoology, meteorology, and surveying.[40]

The fact that by the early twentieth century geographers and the BAAS needed reminding would as surely have astonished them.

Whether Creak's address or the fact that the RGS was now more involved than the BAAS with geodetic work was the prompt to the establishment in 1917 of a BAAS Geodetic Committee is unclear. What is clear is that this committee was motivated by those self-same issues of global co-ordination that had underpinned the Magnetic Crusade and sought to justify this via a rhetoric of imperial benefit. 'In my opinion,' noted Col. G. S. Burrard, the Surveyor General of India, 'the creation of a geodetic Institute for the British Empire is of greater primary importance than the formation of an International association. A British Institute, with Geodetic School attached, is the first essential step towards the unification of the English, Indian, African, Australian and Canadian geodetic researches.'[41] Discussions about the imperial network of geodetic stations were soon overshadowed by debate within Britain about the best place for such a Geodetic Institute and dissent amongst BAAS figures over the nature of the science: the committee achieved little. A short-lived sub-committee entitled Geodetic Surveys in the Colonies was instituted in 1933: it re-emphasised the importance of co-ordinated primary survey to the place of science in Britain's colonies but, upon asking for additional revenue for these purposes, was rebuffed by the Prime Minister's office, given that there 'no funds for such survey work outside ordinary revenue purposes'.[42] By 1945 the Geodetic Committee had ceased to function.

A century after its 'peak' of engagement in co-ordinated global science, BAAS interest in terrestrial magnetism and geodesy ended in squabbles over funding and the best of British locations for geophysical work. Seen in wider context this micro-history and local–national–global geography of terrestrial magnetism is important in relation to what else the BAAS was doing and how it was trying to do it, specifically in terms of the place and nature of geography. As it appeared in sectional terms in the emergent association, geography was ineluctably linked to the physical sciences and was so as leading figures within the BAAS advocated magnetic and geodetic studies as, almost, emblematic sciences for a body needing to secure scientific and political status. If, then, the success of the early Association rested in part upon the success of the Magnetic Crusade and it, in turn, upon the expeditionary cultures of polar exploration, heroic endeavour and systematic observation in the field, it is easy to see why geography was central to such scientific discourse as both means and end. Securing geography's place as an exploratory practice in BAAS affairs before about 1845 was the result of Murchison rhetoric and of Herschelian and Sabinian pragmatics. Once secured – through the work of James Clark Ross and others – and with its exploratory importance promoted by Murchison, reiterated by him in presidential announcements and at moments reinforced in public as Markham did in Dublin in 1857, it is easy to see how geography became synonymous in the Association with exploratory traditions and with geodetic and topographic concerns.

Yet this is to present only a partial picture of how the BAAS and geography within it took intellectual and structural shape as a science of the physical world. Geography was part of debates within the BAAS and Victorian science and society more generally because exploration accorded with contemporary views about imperial power, and because observation and systematic measurements were central methodological concerns in mapping and in the promotion of the map as a tool for the (re)presentation of the world's geographical variety.

The imperious authority of mapping

At the 1834 Edinburgh BAAS meeting, Roderick Murchison was a man on a mission. In addition to lending his support to proposals for work in terrestrial magnetism, he addressed his audience on the 'state and progress of physical geography in Scotland, and the North of England'. Whilst concerned in part with such questions of economic utility that might arise from knowing the country rock better, Murchison's paper was less illustrative than remonstrative, attacking the government and specifically the Ordnance Survey over the slow progress of topographic mapping. Following his paper and discussion upon it, a resolution was proposed and a memorial passed from BAAS General Council to the Chancellor of the Exchequer: 'That the progress of various branches of science is seriously retarded by the want of an accurate Map of North Britain and the north of England, and that this opinion be represented to His Majesty's Government, with the view of expediting the completion of the unpublished or unfinished parts of the Ordnance Survey.'[43] For Murchison and the other BAAS memorialists, the slow progress of the Trigonometrical Survey in North Britain was a cause for concern of itself and relatively – because of the good state of progress in Ireland and in southern England. In placing 'the evils of the want of these Maps in a clear light, they venture to put forth the principal grounds upon which this appeal is founded': in addition to scientific advance, the memorialists noted benefits to railway and canal construction, correction of the coastal surveys and the public benefit and spirit of 'local enterprize' attendant upon mapping. Not least amongst the concerns expressed was the matter of comparative national standing: 'In this untoward condition of a national geographical survey, Great Britain stands almost alone among the civilized nations of Europe, whilst it is obvious that in no country can the production of such Maps be more imperiously called for.'[44]

Several points may be made of this connected 'moment': to do with the history of the Ordnance Survey in Britain, Murchison's scientific life, and the emphasis given to mapping as a scientific and national imperative. The mapping of Britain by the Ordnance Survey in this period was indeed distinctly uneven, the result of battles over different scales, underfunding, contradictions over the purpose of the project and the sheer size of the task.[45] Murchison promoted mapping as an essential scientific and political tool throughout his life, but in the 1830s especially he knew it to be a requirement in developing his career as a geologist and as architect for the earth sciences.[46] When, over twenty years after his Edinburgh speech, Murchison was interviewed as an expert witness by a parliamentary committee scrutinising the Ordnance Survey's work in Scotland, he again found fault with the speed of national topographic mapping. Doing so, he stressed the need for accurate maps and differentiated between property maps at different scales for the benefit of landowners and 'real maps' connected 'with geographical purposes'.[47]

This is not to suggest that Murchison's 1834 Edinburgh talk and the BAAS memorial that followed should be seen as the foundational moment for the promotion of mapping by the BAAS, any more (say) than Huxley's 1860 Oxford talk was alone responsible for the promotion of Darwin's ideas, Tyndall in Belfast in 1874 for the decline of religious views in scientific explanation or, to give another example which connects with the physical sciences under discussion here, Louis Agassiz's influential discussions on land ice at the 1840 Glasgow meeting alone shaped 'modern' glacial theory.[48] Rather, Murchison's attention to the authority of the map as a basis to scientific enquiry and its representation was lent emphasis by what else was happening then: the near-contemporary emergence of cartography (the term did not exist before 1839) and its languages of standardisation; the development of formal bodies for national mapping (of which the Ordnance Survey in Britain was one); and the parallel development of new forms of 'statistical graphics', in mapping and in the sciences as a whole, such as Alexander von Humboldt's use of the isoline, which allowed natural processes and geographical distributions to be depicted in symbols.[49] Maps were both precursors and significant consequences of exploration, and exploration was central to Britain's imperial requirements as distant dominions were put to territorial order. In such ways, geography and mapping were almost coterminous practices in the shaping and depiction of empire.[50]

What the evidence for the BAAS shows is how mapping as a representative form of science, a depictive means, provided geography with a graphic language and an epistemological legitimacy – and, of course, provided a way in which geographer-explorers could direct their endeavours and, once returned, illustrate their printed accounts and public talks (Figures 5.3–5.4). This is apparent in the topics for which BAAS grants were awarded and in many of the presidential addresses. It is clear, too, in what we might think of as those 'hidden' or unrealised emphases behind BAAS geography (and science), namely the 'Recommendations', as they were known. As appeals for financial support or for the formation of a committee on a given topic they took one of two basic forms: most commonly, from a BAAS section to the General Council, and, of a few, from Council to government or associated bodies. Of the recommendations made between 1852 and 1900, thirty-five were forwarded by section E to BAAS General Council. Of this total, sixteen were explicitly on mapping and survey. Six were on the importance of polar exploration (four stressing in one way or another 'the importance of the scientific results to be obtained from Arctic Exploration', as was noted in 1873, with two on Antarctic work being forwarded in 1886 and 1894, the second of which helped to pave the way for Markham's

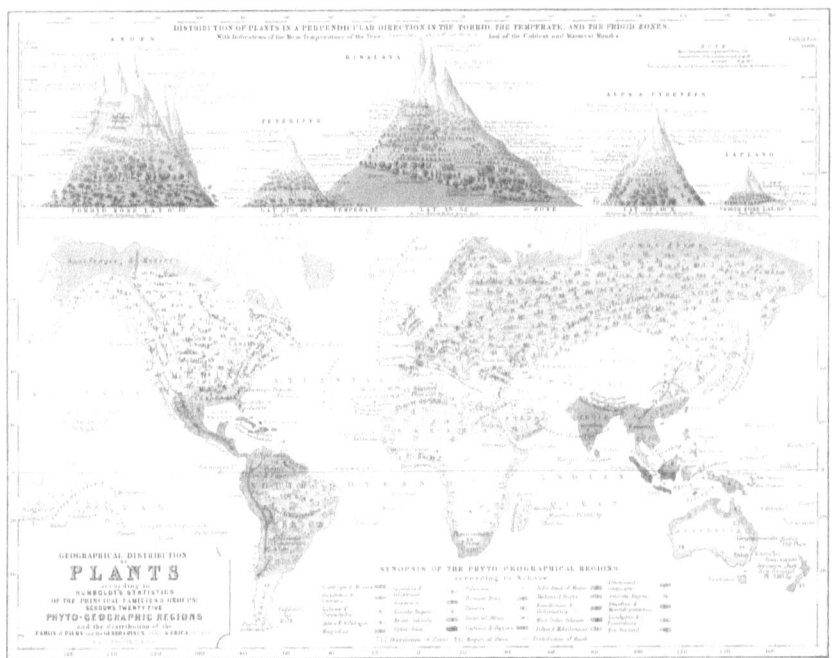

Figure 5.3 The geographical distribution of plants, part of new forms of graphical depictions of natural relationships

formal propositions at the 1895 Ipswich BAAS meeting and at the International Geographical Congress that year regarding the renewal of Antarctic exploration). Four recommendations were made about the scientific significance of hydrography and oceanography, a further three on the Niger, two on geography and education, and one each on the White Nile, India, the geographical distribution of domestic animals and one, with section A, on the standardisation of time (itself part of international debates on the fixing of Greenwich as the 0° meridian). Put another way, only three recommendations – those on the development of geographical education and the proposed establishment in 1863 of a committee on the geographical distribution of domestic animals (a committee which would have included Alfred Russel Wallace had the recommendation been adopted) – did not involve maps as the final product or mapping as the principal method.[51]

BAAS annual meetings likewise provided a forum for the promotion of map work as a central feature of what geography did. As we have seen with regard to the 1872 BAAS meeting in Brighton, the public came to hear Henry Morton Stanley and learn more of the sensationalist

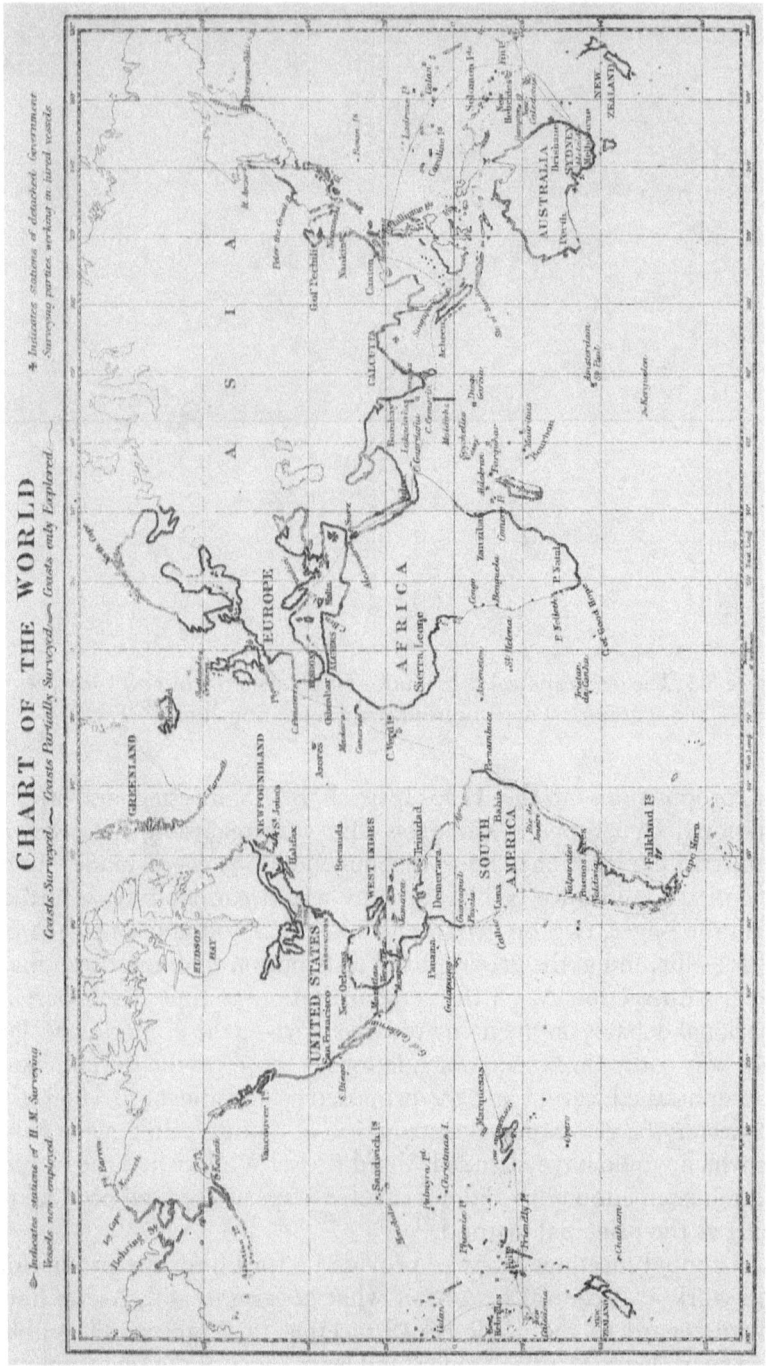

Figure 5.4 The global reach of British hydrographic surveying in 1876

nature of African exploration. In contrast, Francis Galton, African traveller, meteorologist and student of heredity, used his presidential speech there to argue that 'the work of exploration was reaching its final limit, and that the future labours of geographers would have to be directed to obtaining a truer knowledge of the effects of soil and climate upon the physical condition of countries'.[52] Galton also used his lecture to propose that Ordnance Survey maps in pocket form should be sold to the public at a moderate price; others outlined the progress of exploration and mapping in Moab, in the Near East; and yet others used the geographers' platform to campaign vociferously for the overall scientific benefits to be gained from renewed Artic exploration: 'The greatest advantages to various branches of science were insisted upon, and the testimony of Dr. Hooker, representing the botanists, of the Geological Society, of Professor Newton on behalf of Zoology, Mr. Glaisher, the Meteorologists, was given in proof.'[53]

Geography's map and survey work was also significantly about politics and empire, not alone topography and empiricism with a view to disciplinary distinctiveness and scientific connections. What Murchison undertook before 1871 in these respects, men like Clements Markham thereafter did: for Murchison in Edinburgh in 1834, read Markham in Sheffield in 1879. At the latter meeting, Markham drew together and chaired a session of eleven papers, seven of which were on the geographical and strategic questions arising from the Second Afghan Wars (1878–80). Speaking to the importance of topographical survey (and the difficulties of mapping under fire) on the North West Frontier, Lieutenant General Sir Henry Thuillier remarked 'that they might congratulate themselves and geographers generally upon the admirable amount of information they had gained through the instrument of war in Afghanistan': 'Their excellent officers,' he continued, 'had to fight their way in the morning, and get geography [done] in the afternoon.'[54] Following this paper and one by J. O. N. James of the Indian Survey, Markham summarised the mapping work described: as the local newspaper put it:

> The PRESIDENT expressed the opinion that the scientific surveyors were doing a far more valuable and quite as difficult and dangerous work as the explorers who excited more general attention. (Hear, hear.) He had no hesitation in saying that the work of surveying India was the grandest and most glorious work that Great Britain had been employed in during this century. It was impossible for England [sic] to govern India well and faithfully unless she possessed accurate statistical information with respect to the country.[55]

Markham practised what he preached. His paper, published that year, on the mountain passes of the Afghan frontier, based partly on his India

Office experience, stressed the connections between mapping, scientific knowledge and the reach of empire: the 'extraordinary historical interest' attaching to this mountain range 'enhances the importance of systematising and arranging the geographical knowledge connected with it':

> The range also presents certain peculiarities from its position, which makes its examination specially desirable for the furtherance of several branches of scientific investigation in their relations to geography. For instance, as regards botany, the exploration of the Sulamani Mountains will tend to show the relative distribution of members of the Persian and Indian floras which lie on either side of them; and the same interest attached, for similar reasons, to researches in zoology.[56]

And he underscored his view of the key scientific and political role for survey work and topographical mapping unequivocally in his presidential address:

> Our first work as geographers is to measure all parts of earth and sea, to ascertain the relative positions of all places upon the surface of the globe, and to delineate the varied features of that surface. . . .
>
> Accurate maps are the basis of all inquiry conducted on scientific principles. Without them a geological survey is impossible; nor can botany, zoology, or ethnology be viewed in their broader aspects, unless considerations of locality, altitude, and latitude are kept in view. Not only as the basis of scientific inquiry, but also for the comprehension of history, for operations of war, for administrative purposes, and for the illustration of statistics, the uses of accurate maps are almost infinite.[57]

Even, then, as the popular acclaim afforded explorers' accounts, in person and in prose, was giving way to more measured admiration of trained surveyors and scientific geodesists, geography's credentials depended upon a new division of scientific labour and one at work differently for different parts of the world. For Markham, 'In the division of labour, the geodesist produces the accurate large-scale maps which are necessary in thickly populated countries, the topographical surveyor furnishes less exact maps of more thinly peopled and less civilised regions, while the trained explorer forces his way into the unknown parts of the earth.'[58] As modern scholars have pointed out, topographic mapping on the edge of empire involved indigenous experts in the field just as much as it involved skilled geodesists and armies of surveyors. Efforts in fieldwork were anyway only ever 'made real' once the information returned home safely and was put into standard order: in museum, botanical or zoological catalogue, in map rooms and on maps themselves as inscriptions designed to figure the earth.[59] The explorer need not return to be feted: consider Franklin in this respect, or Livingstone. For many involved with the BAAS throughout the nineteenth century, however, mapping was geography's end in view and a defining feature of geography's place in the sciences.

Conclusion

In a wide-ranging address to the Glasgow meeting in 1901, Hugh Robert Mill looked back on a century in which exploration – 'undoubtedly the first duty of geographers' – was, nevertheless, a 'duty which has been well done'. For Mill, geography must move on: 'It must pass beyond the stage of a recreation for retired officers, colonial officials, and persons of leisure, and become the object of intense whole-hearted and original study by men of no less ability who are willing to devote, not their leisure, but their whole time to the work.' Mill even proposed a working definition and invited discussion upon it: 'Geography is the science which deals with the forms of relief of the earth's crust, and with the influence which these forms exercise on the distribution of all other phenomena.' He went on to offer a classification of geography in terms of mathematical geography, physical geography and bio-geography – 'the geographical distribution of life, [which] arises directly from physical geography' – and anthropo-geography, which 'detaches itself from the rest of Bio-geography by the number of exceptions to general laws of distribution and by the human power of modifying the environment'. The object of geographical research, reasoned Mill, 'should be nothing less than the demonstration or refutation of what we claim to be the central principle of geography – that the forms of terrestrial relief control all mobile distributions'.[60]

In his discussion of Murchison's political oversight of geography and geology as sciences of the physical world, a story geared more to his role within the RGS than within the BAAS, Stafford notes how 'the Humboldtian concept of geography to which Murchison adhered, stressing exploration and delineation of the physical landscape rather than the history of human environmental adaptation, provided a set of working parameters for much useful research'.[61] What this chapter has revealed is how much this concept of geography was worked at and worked out in relation to different debates and practices within the BAAS: in regard to terrestrial magnetism; the importance of geography as a science of distribution against which other emergent scientific subjects might place their work; the role of exploration in the polar regions, in Africa and in the tropical regions; geography's relative hierarchical status in relation to botany, geology and the mathematical sciences; and the authority placed upon topographical mapping and geodetic survey, especially by the end of the nineteenth century.

In these terms, geography was seen, in and from the mid 1830s, as a science of exploration – concerned with the location of phenomena, with space and what it contained. Exploration and its attendant mapping was directed more at the form than at the content of the world. By the mid and later nineteenth century, geography as a science of the physical world was more concerned with the relationships between natural things in those spaces, with content and with distributional and causal connections, and with their accurate depiction in map form. This description should not polarise too sharply: even as Murchison and Sabine and others were promoting geography's exploratory credentials in 1834 and 1835, zoologists and not just botanists within the newly formed natural history section were undertaking work on geographical distribution as a feature of their developing sciences.[62] What, later in the century, Owen and Hooker did in emphasising distributional questions as one of the elements marking their science Murchison recognised in 1846 was taking place more generally before then: 'Whilst investigations into the geographical distribution of animals and plants have occupied a large share of the attention of our Browns and our Darwins, it is pleasing to see that some members, chiefly connected with physical researches, are now bringing these data of natural history to bear upon climatology and physical geography.'[63] It is of interest, too, that as the gentlemen of science came together to mark the death of Alexander von Humboldt in 1859 by proposing the establishment of the 'Humboldt Foundation for Physical Science and Travel' – this topic being discussed by BAAS practitioners at its 1859 Aberdeen meeting and again at Oxford in 1860 – Humboldt was revered as a scientist by the Royal Society in letters from Edward Sabine, from the RGS in correspondence from the then secretary, Norton Shaw, and by Murchison, who spoke of the 'truly illustrious Humboldt', on behalf of the BAAS.[64] All sections made a contribution to the funds being co-ordinated by Sabine: among the eighty-four names listed as individual subscribers were James Clark Ross, Murchison, Charles Lyell, David Brewster and Richard Owen.[65]

My point is that whilst we can see 'geography emergent' as a physical science in the context of science's development from the 1830s, it did not emerge in ways that left clear boundaries between it and the other developing disciplines. Nor did it depend upon practitioners who defined themselves (even if they were, perhaps, seen as such by others) as professional geographers. In the same way, contemporaries did not laud Humboldt as a geographer – for what, precisely, was that? – but as a natural scientist. Rather, geography emerged as a practice, even a suite of practices – embracing observation, exploration, of distributional representation, latterly of specialised cartographic endeavour with a view to regional description – which were, in part, shared by some other emergent sciences even as they came to be associated with what geography in particular was. It is noteworthy, too, that whilst polar work helped place geography within the initial remit and hierarchy of the BAAS, that branch of exploration within geography did not again feature strongly until the end of the nineteenth century. For Douglas Galton, speaking in 1895 and reviewing the comparative position of geography between the mid 1830s and the mid 1890s as part of an overview of the sciences, geography as a physical science had helped accomplish much in the polar realms – but much remained to be accomplished:

> Yet the first decade of the British Association was coincident with a considerable development of geographical research. The association was persistent in pressing on the Government the scientific importance of sending the expedition of [James Clark] Ross to the Antarctic and of Franklin to the Arctic regions. We may trust we are approaching a solution of the geography of the North Pole; but the Antarctic regions still present a field for the researches of the meteorologist, the geologist, the biologist, and the magnetic observer which the recent voyage of M. Borchgrevink [the Norwegian polar explorer] leads us to hope may not long remain unexplored.[66]

Alongside geography's place as a physical science, we have also to determine how geography dealt with the human world: there, too, extant disciplinary histories founder in the face of complex connections between people, subjects and the purpose of geographical enquiry.

Notes

1 My use of this term, expressive of uneven process, contingency and situated social context, is intended to contrast with Joseph Conrad's use of the terms 'Geography Militant' and 'Geography Triumphant' by which he described, especially for the nineteenth and early twentieth centuries, the almost predetermined rise of geography as the science of exploration and global (British) dominion. See F. Driver, *Geography Militant: Cultures of Exploration and Empire* (Oxford: Blackwell, 2001), pp. 3–6.
2 C. Daubeny, 'Address by Professor Daubeny', *Report of the Sixth Meeting of the British Association for the Advancement of Science* (London: John Murray, 1837), pp. xxi–xxxvi, quote at p. xxiii.
3 On these points, see J. Endersby, 'Classifying sciences: systematics and status in mid-Victorian natural history', in M. Daunton (ed.), *The Organisation of Knowledge in Victorian Britain* (Oxford: Oxford University Press in association with the British Academy, 2005), pp. 61–86, especially pp. 62–6;

and, at greater length, J. Endersby, *Imperial Nature: Joseph Hooker and the Practices of Victorian Science* (Chicago: University of Chicago Press, 2008), pp. 36–41, 229–32, 234–5, 238–42, 245–8.

4 R. I. Murchison, 'Presidential address to the British Association for the Advancement of Science', *Report of the Sixteenth Meeting of the British Association for the Advancement of Science* (London: John Murray, 1847), pp. xxvii–xliii, quote at p. xxxiii.

5 G. Dickie, *Flora Abredonensis, Comprehending a List of the Flowering Plants and Ferns found in the Neighbourhood of Aberdeen* (Aberdeen: no publisher, 1838), p. i; Endersby, *Imperial Nature*.

6 This point is made by Godlewska in her treatment of Humboldt's botanical work, and by Dettelbach in examining Humboldt's 'global physics': A. M. C. Godlewska, *Geography Unbound: French Geographic Science from Cassini to Humboldt* (Chicago: University of Chicago Press, 1999); M. Dettelbach, 'Global physics and aesthetic empire: Humboldt's physical portrait of the tropics', in D. P. Miller and P. H. Reill (eds), *Visions of Empire: Voyages, Botany, and Representations of Nature* (Cambridge: Cambridge University Press, 1996), pp. 258–92; see also J. Browne, *The Secular Ark: Studies in the History of Biogeography* (New Haven CT and London: Yale University Press, 1983).

7 Cited in Enderby, 'Classifying sciences', p. 65.

8 R. I. Murchison, 'Presidential address', *Report of the Seventh Meeting of the British Association for the Advancement of Science* (London: John Murray, 1839), pp. xxxi–xliv, quote at p. xl.

9 R. I. Murchison, 'Geography and ethnology', *Report of the Thirtieth Meeting of the British Association for the Advancement of Science* (London: John Murray, 1861), pp. 148–53, quote at p. 148.

10 O. J. R. Howarth, 'The centenary of Section E (Geography)', *Advancement of Science* 8 (1951), pp. 151–65, quote at p. 152. On Murchison's formative role more generally, see R. Stafford, *Scientist of Empire: Sir Roderick Murchison: Scientific Exploration and Victorian Imperialism* (Cambridge: Cambridge University Press, 1989).

11 Cited in Mill, 'Geography at the British Association', p. 337, original emphasis.

12 M. Somerville, *Physical Geography* (London: John Murray, 1848), p. 1.

13 Endersby, *Imperial Nature*, pp. 244–5; Richard Owen, 'Address', *Report of the Twenty-eighth Meeting of the British Association for the Advancement of Science* (London: John Murray, 1859), pp. xlix–cx, quote at p. lxxix.

14 Recommendations made at the 1835 Dublin meeting stressed 'the purpose of making observations and discoveries in various branches of Science, as Geography, Hydrography, Natural History, and especially Magnetism, with a view to determining precisely the place of the Southern Magnetic Pole or Poles and the direction and intensity of the magnetic force in those regions': 'Resolutions of the Committee of Magnetism', *Report of the Fifth Meeting*

of the British Association for the Advancement of Science (John Murray, London, 1835), p. xxi.
15 *Memorial of the British Association for the Advancement of Science, in relation to the Present State of Survey of North Britain*, British Parliamentary Papers 1836 XLVII, pp. 89–96.
16 For discussions of this topic, see J. Cawood, 'Terrestrial magnetism and the development of international collaboration in the early nineteenth century', *Annals of Science* 34 (1977), pp. 551–87; J. Cawood, 'The Magnetic Crusade: science and politics in early Victorian Britain', *Isis* 70 (1979), pp. 493–518; J. Morrell and A. Thackray, *Gentlemen of Science: Early Years of the British Association for the Advancement of Science* (Oxford: Clarendon, 1981), pp. 353–70, 523–31; R. T. Merrill, *The Earth's Magnetic Field: Its History, Origin and Planetary Perspective* (London: Academic Press, 1983). G. A. Mawer, *South by Northwest: The Magnetic Crusade and the Contest for Antarctica* (Edinburgh: Birlinn, 2006). I am grateful to Dr H. Lewis-Jones of the University of Cambridge for allowing me access to his unpublished paper 'Attractive Visions and an Imagined Community: Terrestrial Magnetism, Rhetoric and the British Association for the Advancement of Science'.
17 Stephen Pumfrey, *Latitude and the Magnetic Earth* (Cambridge: Icon Books, 2002).
18 On Enlightenment interest in terrestrial magnetism and the 'shape of the earth' debate, see R. M. Yost, 'Pondering the imponderable: John Robison and magnetic theory in Britain, *c*. 1775–1805', *Annals of Science* 56 (1999), pp. 143–74; R. Iliffe, '"Aplatisseur du monde et de Cassini": Maupertuis, precision measurement, and the shape of the earth in the 1730s', *History of Science* 31 (1993), pp. 335–75; M. Terrall, *The Man who Flattened the Earth: Maupertuis and the Sciences in the Enlightenment* (Chicago: University of Chicago Press, 2003).
19 Cawood, 'The Magnetic Crusade', p. 494.
20 [Anon.], 'Resolutions of the Committee of Magnetism', *Report of the Fifth Meeting of the British Association for the Advancement of Science* (London: John Murray, 1836), pp. 1–34, quote at p. 21.
21 E. Sabine, 'Observations on the direction and intensity of the terrestrial magnetic force in Scotland', *Report of the Sixth Meeting of the British Association for the Advancement of Science* (London: John Murray, 1837), pp. 97–119.
22 Cawood, 'The Magnetic Crusade', p. 509. On naval men and their scientific role in the Royal Navy, see Randolph Cock, 'Scientific servicemen in the Royal Navy and the professionalisation of science, 1816–1855', in D. Knight and M. Eddy (eds), *Science and Beliefs: From Natural Philosophy to Natural Science, 1700–1900* (Aldershot: Ashgate, 2005), pp. 95–111.
23 Mawer, *South by Northwest*, passim.
24 F. Fleming, *Barrow's Boys* (London: Granta, 1998).
25 J. Herschel (ed.), *The Admiralty Manual of Scientific Enquiry* (London:

John Murray, 1849), p. iii; R. Yeo, *Science in the Public Sphere: Natural Knowledge in British Culture 1800–1860* (Aldershot: Ashgate, 2001), p. 266.
26 S. F. Cannon, *Science in Culture: The Early Victorian Period* (New York: Dawson, 1978).
27 J. Cavell, *Tracing the Connected Narrative: Arctic Exploration in British Print Culture, 1818–1860* (Toronto: University of Toronto Press, 2008).
28 [Anon.], 'Resolutions of the Committee of Magnetism', *Report of the Fifth Meeting of the British Association for the Advancement of Science* (London: John Murray, 1836), p. 21.
29 Mawer, *South by Northwest*, pp. 39–44, 58–60, 66, 81–2, 86, 93–7, 100, 104–6.
30 R. I. Murchison and E. Sabine, 'Address', *Report of the Tenth Meeting of the British Association for the Advancement of Science* (London: John Murray, 1841), pp. xxxv–xlviii.
31 J. Herschel, 'Address', *Report of the Fifteenth Meeting of the British Association for the Advancement of Science* (London: John Murray, 1846), pp. xxvii–xliv, quote at p. xxxiv.
32 R. Murchison, 'Address', *Report of the Sixteenth Meeting of the British Association for the Advancement of Science* (London: John Murray, 1847), pp. xxvii–xliii, quote at p. xxviii.
33 R. Stafford, *Scientist of Empire: Sir Roderick Murchison, Scientific Exploration and Victorian Imperialism* (Cambridge: Cambridge University Press, 1989), pp. 223–4, *passim*; J. Secord, 'King of Siluria: Roderick Murchison and the imperial theme in nineteenth-century British geology', *Victorian Studies* 25 (1982), pp. 413–42; F. Driver, *Geography Militant: Cultures of Exploration and Empire* (Oxford: Blackwell, 2001), pp. 38–9.
34 Cawood, 'The Magnetic Crusade', pp. 516–18.
35 M. S. Reidy, *Tides of History: Ocean Science and Her Majesty's Navy* (Chicago: University of Chicago Press, 2008), especially pp. 281–4. Reidy's work significantly develops that of Morrell and Thackray on this topic, a subject they treat briefly as 'Tidology': Morrell and Thackray, *Gentlemen of Science*, pp. 513–17.
36 Royal Geographical Society Archives, CRM 74, fols 1–1v, 28 August 1857. Markham's paper was published in the Dublin papers the following day.
37 [H. Lloyd], 'Report of the Council of the British Association as presented to the General Committee at Leeds, September 22nd, 1858', *Report of the Twenty-eighth Meeting of the British Association for the Advancement of Science* (London: John Murray, 1859), pp. xxx–xxxi.
38 Royal Geographical Society Archives, CB 1882–1902, Sir Erasmus Ommaney to H. W. Bates, 10 May 1887.
39 Royal Geographical Society Archives, CB 1882–1902, Sir Erasmus Ommaney to H. W. Bates, 4 November 1887.
40 E. Creak, 'Section E: Geography', *Report of the Seventy-third Meeting*

of the British Association for the Advancement of Science (London: John Murray, 1904), pp. 701–11.
41 BAAS Archives, Dep. BAAS 331, Geodetic Committee, 1917–1945, fols 86–7, 28 March 1917.
42 BAAS Archives, Dep. BAAS 333, 'Geodetic Surveys in the Colonies, 1933', fol. 324, 7 October 1933.
43 [Anon.], *Memorial of the British Association for the Advancement of Science, in relation to the Present State of Survey of North Britain*, British Parliamentary Papers XLVII (1836), pp. 89–96, quote at p. 89.
44 *Memorial of the British Association for the Advancement of Science, in Relation to the Present State of Survey of North Britain*, pp. 89, 90.
45 W. A. Seymour (ed.), *A History of the Ordnance Survey* (Folkestone: Dawson, 1978); C. W. J. Withers, 'Authorizing landscape: 'authority', naming and the Ordnance Survey's mapping of the Scottish Highlands in the nineteenth century', *Journal of Historical Geography* 26 (2000), pp. 532–54.
46 Stafford, *Scientist of Empire*, passim.
47 *Report from the Select Committee appointed to consider the Ordnance Survey of Scotland; together with the Proceedings of the Committee, Minutes of Evidence, Appendix and Index*, British Parliamentary Papers XIV (1856), pp. 75–80, quote at p. 80.
48 D. A. Finnegan, 'The work of ice: glacial theory and scientific culture in early Victorian Edinburgh', *British Journal for the History of Science* 37 (2004), pp. 29–52.
49 M. Edney, 'Cartography without "progress": reinterpreting the nature and historical development of mapmaking', *Cartographica* 30 (1993), pp. 54–68; M. Edney, 'Cartography: disciplinary history', in G. A. Good (ed.), *Sciences of the Earth: An Encyclopedia of Events, People, and Phenomena* (New York and London: Garland, 1998), Vol. 1, pp. 81–5; A. M. C. Godlewska, 'From Enlightenment vision to modern science? Humboldt's visual thinking', in D. N. Livingstone and C. W. J. Withers (eds), *Geography and Enlightenment* (Chicago: University of Chicago Press, 1999), pp. 236–79; M. Friendly and D. Denis, 'The roots and branches of statistical graphics', *Journal de la Société française de statistique* 141 (2000), pp. 51–60.
50 M. Edney, *Mapping an Empire: The Geographical Construction of British India, 1765–1843* (Chicago: University of Chicago Press, 1997); D. G. Burnett, *Masters of all they Surveyed: Exploration, Geography, and a British El Dorado* (Chicago: University of Chicago Press, 2000); T. Winichakul, *Siam Mapped: A History of the Geo-body of a Nation* (Honolulu HI: University of Hawaii Press, 1994).
51 These numbers are derived from assessment of statements regarding Recommendations from the *Reports* between 1852 and 1900.
52 MS. Dep. BAAS 412, *Brighton Daily News*, 16 August 1872.
53 MS. Dep. BAAS 412, *Brighton Daily News*, 21 August 1872.

54 MS. Dep. BAAS 415, *Sheffield Daily Telegraph*, 21 August 1879, p. 3.
55 MS. Dep. BAAS 415, *Sheffield Daily Telegraph*, 21 August 1879, p. 3.
56 C. R. Markham, 'The mountain passes of the Afghan frontier of British India', *Proceedings of the Royal Geographical Society and Monthly Record of Geography*, 1 (1879), pp. 38–62, quote at p. 41.
57 C. R. Markham, 'Geography', *Report of the Forty-eighth Meeting of the British Association for the Advancement of Science* (London: John Murray, 1879), pp. 420–33, quote at pp. 420, 421. On the British view of Tibet and the archiving of imperial geographical knowledge in this period, see T. Richards, *The Imperial Archive: Knowledge and the Fantasy of Empire* (London: Verso, 1993).
58 Markham, 'Geography', p. 420.
59 See for example Jane Camerini, 'Remains of the day: early Victorians in the field', in B. Lightman (ed.) *Victorian Science in Context* (Chicago: University of Chicago Press, 1997), pp. 354–77; Jane Camerini, 'Wallace in the field', *Osiris* 11 (1996), pp. 44–65; H. Ritvo, 'Zoological nomenclature and the empire of Victorian science', in Lightman (ed.), *Victorian Science in Context*, pp. 334–53; K. Raj, 'When human travellers become instruments: the Indo-British exploration of Central Asia in the nineteenth century', in M.-N. Bourguet, C. Licoppe and H. O. Sibum (eds), *Instruments, Travel and Science: Itineraries of Precision from the Seventeenth to the Twentieth Century* (London: Routledge, 2002), pp. 156–88.
60 H. R. Mill, 'On research in geographical science', *Report of the Seventy-first Meeting of the British Association for the Advancement of Science* (London: John Murray, 1901), pp. 698–714, quotes at pp. 699, 701, 705, 707.
61 Stafford, *Scientist of Empire*, p. 220.
62 In 1834, for example, James Wilson undertook to produce a report 'on the geographical distribution of Insects, particularly *Coleoptra*': *Report of the Fourth Meeting of the British Association for the Advancement of Science* (London: John Murray, 1834), p. xxxvii.
63 Murchison, 'Presidential Address to the British Association' (1846), p. xxxv.
64 MS. Dep. BAAS 405.
65 MS. Dep. BAAS 5, fols 231–3.
66 D. Galton, 'Address', *Report of the Sixty-fifth Meeting of the British Association for the Advancement of Science* (London: John Murray, 1895), pp. 3–35, quote at p. 11.

6

Measurement, exploration and ethnology: geography and the human sciences

This chapter elaborates upon arguments that we have seen for geography as a topographic and cartographic science concerning geography as a science of distribution, the operations of the Association and of section E and the development of its scientific status and agenda. My concern here is with the place of geography, ethnology and anthropology within the BAAS, in terms of subject content, sectional relationships and scientific status. This is, but only in part, a story of human geography, recognising that that term was a feature of BAAS geographical science but notably so only from the end of the nineteenth century. It is at greater length an account of the geography of humans and of schemes over how to 'read', measure and study humans in place, in their different geographical settings. There are, of course, other connections with the foregoing chapter. Key individuals, not least Sir Roderick Murchison, were involved in establishing the agenda for the emergent human sciences, and their institutional place, in the BAAS as they were for the physical sciences. Both physical and human involved the sciences of survey and of distribution: the first, as we have seen, in relation to questions of accurate geodetic measurement, national–international rivalry and collaboration over terrestrial magnetism and an emphasis upon topographic and cartographic work, the second, as we shall see, with respect to ethnographic practice within the emergent human sciences. In the case of terrestrial magnetism and geodetic survey, the politics of associational science centred upon the involvement of leading physicists and mathematicians and their associates, including naval polar explorers, and upon institutional links between the British Association, the Royal Society and the Admiralty and, later in the nineteenth century, between national governments and scientific bodies such as the Royal Geographical Society as geodetic work acquired national and imperial significance and required global co-ordination and management. For persons interested in those questions of human cultural difference and human interaction with nature which underlay the development of anthropology/ethnology and human geography in the nineteenth and early twentieth centuries, the politics of science took the form of heated debates – in spoken addresses, published papers, even in sectional minute books – about scientific focus, matters of practice and sectional identity. Concerns from practitioners within and outwith the BAAS – evident in debates over the establishment of systematic procedures for the more scientific investigation of humans – led to the appearance of 'how to' guides to ethnological fieldwork whose purpose it was to establish epistemological procedures for the doing of a human-centred geography, at home and in relation to the diverse peoples of the British empire.

This chapter shows how, through its component sciences, epistemic procedures and institutional workings, the Association measured up: as a body designed to advance science and in terms of focus and practice culturally, and even physiognomically, in regard to races and peoples. In demonstrating the close institutional connections between the BAAS, the RGS and other bodies in developing the shape of BAAS civic science as a human-oriented science, we can, I suggest, see the science of ethnology as part of the 'institutional reach' of the BAAS as it subjected the subject peoples of the British Empire and some of the indigenous peoples of Britain to empirical scrutiny. What is also revealed is something of the struggle for intellectual authority between and within proponents of ethnology, anthropology and geography over who would do such work, and how. When a more evidently human geography started to appear at the end of the nineteenth century, it did so, I contend, more outside the BAAS than within it. To see why this is so and to understand the place of human geography and of the human sciences within the BAAS demands that we stand back from the BAAS.

Geography, ethnology and anthropology c. 1839–c. 1935: epistemology and identity in the human sciences

It is not easy to make sense of geography's place as a human science in nineteenth-century Britain, or, rather, of its place in respect of the BAAS. The specific context was much shaped by debates between proponents of what would become ethnology and anthropology and by the foundation of separate societies for these sciences before, in 1871, differences were settled and the two bodies merged. More generally, geography's presence in the BAAS and in the complex debates from the 1830s over human cultural characteristics, human variability and racial origins was part of closely intertwined matters then giving shape to science (and to the BAAS) as an intellectual and political resource: the development of institutional and disciplinary spaces, the emergence of distinguishing professional

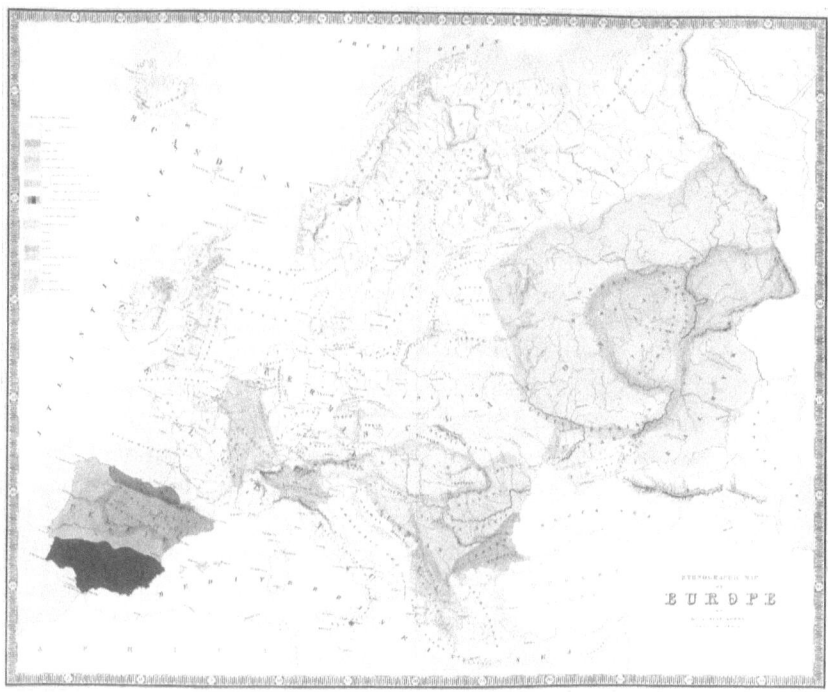

Figure 6.1 Ethnographic map of Europe in 1850, by the German ethnographer Gustav Kombst

practices, heated debates about the deepening of the earth's history in relation to scriptural interpretation, and, not least, disputes over the origins and immutability (or not) of humans as a species. Such issues have been covered *in extenso* by others.[1] Furthermore, for many contemporaries addressing such matters, 'geography' was commonly equated with 'physical geography' not just in the senses we have recorded above but in the sense of the environmental setting in which human and non-human objects occurred. So understood, physical geography provided a means wherein human differences might be explained – as 'determined' by that environment. As noted, one result was that as a method of interpreting and presenting their work, many nineteenth-century writers upon earth history, nature's variability and human difference saw geography and cartography as essentially synonymous. As science became more secure in its methods and over its objects in view, so atlases provided a key illustrative space for the interpretation of geography as nature's phenomena in space and texts seeking greater objectivity in the representation of nature, even if mankind in its diversity did not replicate on the ground the neat delimitations of cartographers[2] (Figure 6.1).

Whilst acknowledging these issues, my concerns are more modest and more precise. In looking at the sectional place of geography, ethnology and anthropology and at their intellectual connections, initially between 1839 and 1871, I want to reveal how the study of humans fell – and did not fall – within the orbit of section E and so, in turn, cast light upon the shaping of the human sciences within the BAAS. This is not just and it is certainly not simply a matter of sectional history. Geography as a human science was shaped by personal conflicts and epistemological debates between ethnologists and 'anthropologicals' over the nature of human history even as it provided, through exploratory encounter with different peoples, the raw material for such debates. By considering geography's sectional and scientific history in closer detail we can illuminate the wider context in which geography as a BAAS human science was set. This too is a story of emergent practice. Since students of the human world did not have agreed procedures for systematic study, it is understandable that practitioners should move to establish guidelines to practice. Addressing geography's making in these ways is to make clear why different human groups and human cultures at home and abroad became the object of geographical attention from within the BAAS in the ways they did.

Human subjects and scientific identity: geography, ethnology and anthropology, 1839–1871

Section E's designation from 1851 as 'Geography and Ethnology' did not signal the start for study of the human sciences within the BAAS. Formally, matters to do with the human world were initially taken up by natural history – what would in time be section D, biology. If there is a foundational moment to geography's human face, as it were, it was in Birmingham in 1839, in a city then shaken by radical politics and suffering economic depression. For the first time, an emergent 'ethnological agenda' was included within the meeting, led by the anti-slavery campaigner Thomas Hodgkin and the ethnologist James Cowles Prichard, author of *Researches into the Physical History of Mankind* (published at various dates between 1826 and 1847). Talking to the title 'On the Extinction of some Varieties of the Human Race', Prichard addressed the BAAS meeting on the potential loss to science that would follow if colonisation were to lead to aboriginal extinction. For the thirty-year-old Charles Darwin present in the audience, Prichard's and Hodgkin's talks affirmed his profound anti-slavery sentiments and helped shaped his ideas on humans' common ancestry.[3]

For students of the human world within and beyond the BAAS, there were no secure methodological guides over how to investigate human diversity. How should one collect, consistently, reliable data on human physical variety? On human customs? Fieldwork and expeditions allowed direct engagement – as Darwin could testify of his *Beagle* voyage and his encounters with Fuegians, Patagonians and Maoris, amongst others – just as we have seen how, for Murchison, Herschel, Sabine and others, Ross's *Erebus* and *Terror* Antarctic voyages provided ample proof of the value of expeditionary science in measuring physical phenomena. Recognising that, for most students of human variety, scientific knowledge might come from expeditionary contact but that, when and where it did, it required regulating and trusting others, Prichard and delegates at the Birmingham natural history meeting drew up and printed *Queries respecting the Human Race, to be addressed to Travellers and others*, in order to guide ethnological procedures.

This was no simple questionnaire. Paralleling collaborative initiatives in magnetic science, the 1839 *Queries* were drawn up in part from comparable guides being produced by the Ethnological Society of Paris as well as by Prichard and others. As with terrestrial magnetism, here was an international dimension to the development of scientific practice in the early BAAS. Even so, what underlay the desire for method was the political purpose for science in terms of Britain's global reputation and imperial responsibilities:

> Britain, in her extensive colonial possessions and commerce, and in the number and intelligence of her naval officers, possesses unrivalled facilities for the elucidation of the whole subject; and it would be a stain on her character, as well as a loss to humanity, were she to allow herself to be left behind by other nations in this inquiry.[4]

Queries brought together dozens of questions, ranging from instructions over physical dimensions to matters of cultural practice. The first question, for example, asked informants to 'State the general stature of the people, and confirm this by some actual measurements. Measurement may be applied to absolute height, and also to proportions, to be referred to in subsequent queries.' Questions were asked about 'the prevailing complexion', about the head particularly – 'The head is so important as distinctive of race, that particular attention must be paid to it' – and, relatedly, about the skull: 'State whether the bones of the skull are thick, thin, heavy, or light.' Other queries focused upon customs. 'Do the natives speak a language already known to philologists, and if so, state what it is.' 'Does infanticide occur to any considerable extent?' 'What is the form of government?' 'Is the religion of the people similar to that of any other people, neighbouring or remote?' And so on.

In the context of the early BAAS, *Queries* was both scientific instrument and procedural manual. It was a preliminary textual device for the interrogation and measurement of human specimens by regulating observers in the field – travellers, credible 'others', British naval officers – as reliable data-generating recording devices. It may be placed in that longer-run history of 'how to' guides about travel, observation and collection that distinguished the development of scientific method in and from this period. Just as the birth of the 'social fact' from the 1830s was rooted in late seventeenth-century prescriptions on how to travel and observe and in eighteenth-century practices of systematic survey, statistical accountancy and political economy, so *Queries* may be placed in that genealogy of instructive geographical literature, in which Herschel's 1849 *Admiralty Handbook* also figures and in which, for geography, *Hints to Travellers*, first published by the RGS in 1854, has been accorded perhaps undue primacy.[5] The cartography of the human world was likewise part of the rise of statistical graphics in the depiction of scientific phenomena, even if human variety did not lend itself as readily as did natural knowledge to quantitative enumeration.[6] There is a yet further context, however, which helps explain the timing of Prichard's *Queries*, the nature of the questions it posed, and the place of the human sciences within the BAAS.

From the 1830s to the 1870s and even later, the study of humans in relation to their geography – not, yet, anything termed human geography – centred upon the politics of race and human origins.[7] Human science was racial science: disputes between the emergent 'disciplines' of ethnology and anthropology were, broadly, disputes between students of human cultural production on the one hand, and, on the other, students of human physical difference from which differences associated moral and intellectual capacities were presumed to follow. Some thinkers turned to cultural questions, others to physical descriptions, writers often seeking corroboration with the scriptures or complicity with racial oppression and slavery as justification for their views. Those determined to see clues to intellectual capacity and racial difference from physical features drew support for their beliefs from the 'science' of phrenology. From the study of head bumps and skull shape there developed an interest in human measurement as a surrogate indicator of cultural and intellectual capacity: anthropology begat anthropometry begat the 'facts' of racial prejudice. Others, less concerned with physical differences and more with the common origins of humans and with the practices shared by them despite geographical separation and differences in environmental milieu, looked to such things as customs and comparative philology and to their explanation in relation to humans' local setting and geographical distribution. As evolutionary ideas developed, as explorers and geographers encountered diversity in human culture and as some looked to the study of human difference and found multiple origins for humans (polygenism) where others saw evidence of single origins (monogenism), no one could escape the fact that 'ethnology was becoming explosive'.[8]

Given this context, it is easy to see why *Queries* contained questions on physical form and on cultural practice, why Hodgkin and others distributed it to BAAS delegates, why the Ethnological Society, founded in 1843 and rooted in the sentiment of political liberalism and antislavery, provided an intellectual home for the 'Ethnologicals' and not for the 'Anthropologicals' (which latter constituency, disenfranchised with the science and the politics of their fellows, founded their own body, the Anthropological Society, in 1863). The sense that what was becoming the geography of humans within the BAAS was because of what was taking place outside the BAAS is supported by closer examination of the activities and workings of these two bodies before 1871 (the year in which the Ethnological Society and the Anthropological Society merged to form the Royal Anthropological Institute).

It is clear from minutes of the Ethnological Society that members of that body saw the ethnological papers as a significant feature behind the popularity of section E as a whole. Crawfurd's paper in 1847 on the Malay languages and races attracted widespread interest: 'We' (the suite of ethnology papers) 'were second to none' in the 'interest excited' at the 1852 Belfast meeting, noted Richard Cull; many later reports are of like tenor.[9] By the mid 1860s, however, figures within the Anthropological Society were anxious about the presence of that science in the BAAS and in section E. They felt that the popularity of ethnology and the lack of scientific discussion over human differences – the two matters being seen, of course, as directly connected – were undermining serious anthropological work. It is clear too, as we shall see, that the association of geography with popularity and with female audiences meant that practitioners of the human sciences wanted to distance themselves from section E and its affairs.

Whilst we can trace in such matters the disaffection of the anthropologists, and see their not-so-veiled views on the nature of ethnology and of geography, it is not immediately clear why geography should present itself, as a section and as an emergent science, with ethnology during the 1850s and 1860s – until, that is, account is again taken of the role of Murchison. Recall Howarth's words that 'It is impossible to overrate the personal power of Murchison in the Section from 1851 to 1870.'[10] The sense that Murchison was busy shaping not just the physical face of

Figure 6.2 Cartoon of Sir Roderick Murchison chairing the BAAS meeting, Birmingham, 1865. In part, the accompanying text reads: '*Geography*. The question "Where are you going on Sunday?" satisfactorily answered. This included practical demonstrations in Street Gymnastics, or the Use of the Globes and Poles'

geography within the BAAS and the RGS but also its human content is evident in the cartoon of him chairing the 1865 BAAS section E meeting (Figure 6.2). Murchison's commitment to geography as the science that served empire, and its different peoples, helped secure the profile of the Royal Geographical Society as a body concerned with topographical mapping, primary survey and physical geography rather more than with the betterment of natives within Britain's imperial orbit.[11] Within the BAAS study of humans, the concern of the anthropologists with physical differences mattered even less if the predominant vision being shaped for geography was of a subject aiming at imperial scrutiny of the physical environment and, less importantly, at the civilisation and moral redemption of other peoples and cultures. We can in these terms see how section E from 1851 came to include geography and ethnology – because, in particular, what was being squeezed out was anthropology, and, in general, because Murchison wanted to retain such ethnological elements as helped keep section E popular – even as he elsewhere promoted geography's credentials as a physical and imperialising science through topographical mapping and exploration.

Murchison had first proposed the 'Geography and Ethnology'

section title rather than 'Anthropology' at the Birmingham meeting in 1849. There was an 'Ethnological sub-section', with Prichard ('my eminent friend', as Murchison termed him) as its first chair, from 1846, but its role was much diminished following Prichard's death in 1847. The name of this new enterprise was the subject of discussion at the meeting the following year: in writing to an anonymous correspondent, Murchison in August 1850 noted simply, 'I carried my point of elevating Geography with a distinct Section viz Section E to be called Geography and Ethnology.'[12]

The choice of the sectional name Geography and Ethnology from 1851 was thus neither accidental nor inconsequential. It was consistent with Murchison's orchestration of parallel developments in the RGS, with his emphasis upon geography's physical topographic agenda, with his patronage of certain figures (ethnologists) within the emergent human sciences, with a certain political and intellectual vision for the geographical study of humans as cultural and imperial beings rather than as physical specimens, and with the slow-wrought separation in identity of geography from geology and of anthropology from ethnology. Murchison's reflections on these issues, made in 1858, are instructive:

> In truth, the geographers and travellers who had been previously tacked on to the Geological Section, finding that they had no chance of a patient hearing among their associates the geologists, were dissatisfied; and . . . I felt assured that their separation, followed by this new amalgamation, would be highly conducive to the best interests of each. Then, again, the ethnologists were discontented at having no local habitation, and at being compelled to form at several meetings a sub-section, seeking for a meeting-room where they best could find it.
>
> The union of geography and ethnology was indeed so natural, the subjects of which they severally treat are each so engaging and instructive, as well as popular, that the result has proved most satisfactory. In short, this Section has been, and I trust will continue to be, well-thronged by votaries, who rejoicing in the spirit of foreign research, come here to gather knowledge from the lips and writings of distant voyagers and travelers by sea and by land.[13]

This union of these subjects may not have been 'natural' but it could be made to seem so, of course, by its leading proponent. The fact that more evidently 'anthropological' issues surfaced from time to time suggests that the section did offer an intellectual space for a variety of interpretations of human geographical matters. Speaking to the title 'On the Connexions between Ethnology and Physical Geography' in his 1861 presidential address, John Crawfurd began thus:

> Man will be found savage, barbarous, or civilised, in proportion to the quality of the race to which he belongs, and to the physical character of the country in which his lot has been cast. . . . Mere intemperance of climate, independent of any other obstacle, is sufficient to prevent man from making any advance towards civilisation, and to hold him permanently in the savage state.[14]

An 'Ethnological' like Prichard, Crawfurd the polygenist nevertheless dismissed totally his predecessor's monogenist views.[15] What was at issue was the balance between ethnological description and anthropological science, in and out of section E.

The fact that this paper was read to the Ethnological Society in October 1861 and published two years later in its journal further demonstrates the close institutional affiliations between that body and influential figures in section E in ways we have seen in other contexts for section E and the RGS. As Murchison remarked of Crawfurd's efforts, 'we may reasonably calculate on receiving much sound support from men like these, who can so well connect the sciences we cultivate with many other branches of human knowledge'.[16]

During the 1860s, however, the ethnologists and the anthropologists were coming closer together. Even as Murchison endorsed the connections between ethnology and geography, and as the 'anthropologicals' railed at their minor role, others cautioned about the range of knowledge embraced by these terms and, thus, their seeming imprecision as sciences. Recognising too the 'indefiniteness of aim' that was geography understood as study of the 'leading features of the earth's physiognomy', the educationalist Sir Charles Nicholson spoke to this matter in his presidential address in 1865:

> The same remark applied to ethnology, the indefiniteness of the term having become a source of difficulty. A fastidious critic might find equal objection to the employment of such terms as ethnography, zoography, anthropology, biology and others. Many of these terms are sufficiently elastic not only to include man in his objective relations (in which anatomy and physiology, human as well as comparative, could be embraced), but all the ethical and moral qualities of his nature would become alike objects of contemplation and research.[17]

For Samuel Baker, speaking the following year, the human sciences were allied with geography and dependent upon them: 'Ethnology is a twin sister of geographical science, as the numerous races of human beings (so diverse and inexplicable) that inhabit the various portions of the earth, from the ice-bound regions of the Arctic to the burning deserts of Africa, would have been unknown but for the researches of geographers and explorers.'[18] But ethnology was also a loose term in the eyes of some within section E.

Such pronouncements from those who saw connections between geography and ethnology but who also recognised an imprecision in the nature of ethnology were paralleled by strident voices from within the anthropological-ethnological community. Speaking in 1865, the controversial James Hunt, polygenist thinker and president of the Anthropological Society, was clear where the blame lay:

> The destroying angel who annihilated the Ethnological sub-section was Sir Roderick Murchison. . . . Now, Sir Roderick did not like geography to be submerged in geology, and yet he felt no compunction in submerging ethnographical or ethnological papers amongst geographical ones. We give great honour to him for what he has done for geographical science, but I know of no ethnologist or anthropologist who will thank him for destroying the sub-section of ethnology.[19]

Hunt with others began the move to have a separate anthropological section within the BAAS, one that incorporated the 'ethnologicals', effectively from January 1865. Differences between the ethnologists and the anthropologists over the scientific study of humans were still apparent. But there was also a stronger sense that the human sciences as a whole were not being well served within section E. What was needed was a science of man which could encompass the divisions between ethnologists and anthropologists, and a section that addressed substantive and methodological questions about human differences, cultural and/or physical. As Hunt noted, writing to the BAAS in August 1865:

> All who are at all acquainted with the practical working of Section E, must be fully aware that virtually there has been no real discussion on the science of man in it since its formation. Occasionally there has been an important paper treating of some branch of the science of mankind, but generally that section is so overburdened with geographical papers that those on other subjects have been obliged to be passed over without being read in full or discussed properly.

Hunt was only emphasising in correspondence what he had anyway earlier proclaimed. Addressing the Anthropological Society in January, he had remarked how 'There is much in our science which can never be made popular, and for which the "Ladies Section, E", is hardly the fit place'[20] (cf. Figure 3.5).

Things came to a head at the 1865 Birmingham meeting: it may be that the cartoon of Murchison juggling several balls in the air (Figure 6.2) is a reference to the competing interests on show at the meeting that year. Hunt brought forward a proposal for a separate anthropological section. This was rejected. Counter-proposals were made – one advocating that ethnology should stay with geography and a newly titled 'science of man' should be linked with biology. This too was rejected. Anthropologists ranted at the nature of section E: 'Geography is too unimportant a science to fill an entire section. . . . Whatever may be the future of the Geographical Society [the RGS], we think the British Association will do well to get rid of those semi-sensation or hero-worship exhibitions, which have become too much associated in this country with the word geography.'[21] Proposals made to name section E 'Geography, Ethnology and Anthropology' in a bid to assuage all parties came to naught.[22] With the matters aired, the cracks widened. In 1869 and in 1870 ethnology was, briefly, placed with biology. In the following year, anthropology was for the first time proposed and then recognised as a separate sub-section of section D in the annual meeting. In the chair was Alfred Russel Wallace, who – probably unhelpfully in the minds of proponents then urging disciplinary clarity and, thus, sectional identity – defined anthropology as 'the science which contemplates man under all his varied aspects (as an animal and as a moral and an intellectual being), in his relations to lower organisms, to his fellow-men, and to the universe'.[23] By 1871, anthropology and ethnology, the two by then linked in a new subject body, came together in the BAAS as section H.[24]

In 1870, Murchison, by then frail and attending what was to be his last BAAS meeting (he died in October 1871), reflected further on the schism and the presence, by then, of an anthropological sub-section:

> This Section E had for many years the title of Geography and Ethnology; but the latter term has recently been abstracted from us, Ethnology having been relegated to the newly constituted comprehensive Section of Biology. Now, although I have presided over this Section when it possessed its double title, I admit the value of the change, seeing that we are relieved from the duty of receiving and discussing those anthropological memoirs which are intimately connected with physiology and comparative anatomy. Such memoirs could not be adequately discussed by geographers, and they are now submitted to competent judges.

Even so, Murchison wanted to retain such matters to do with human description and customary beliefs as might help ensure geography's popularity as a section:

> At the same time I earnestly hope that papers relating to Ethnography, including accounts of the language and customs of distant peoples, and which is intimately bound with the physical geography of countries will, as heretofore, flow into our hall, and thus render our meetings on this occasion a successful and popular as they have been during past years.[25]

In short, what ethnologists saw between about 1839 and 1851 as studies in human description and anthropologists as the practice of human measurement – to the anthropologists, the first was popular, fit for a 'Ladies' Section', the latter was good science – was part of a complex struggle for recognition as a BAAS section and, more important, as human science. Outside the BAAS, reconciliation was achieved in time: in the formation in 1871 of the Royal Anthropological Institute. Inside the BAAS, geography as a section and as a science was being shaped by the partial initial presence of ethnological work (where it served Murchison's popularising purpose), by the endorsement of geography's exploratory credentials in as much as they helped reveal the variety of forms of human social existence and, but only latterly, by the separation of anthropology and ethnology as a section. This did not mean that section E was without a human focus after 1871 – within its later meetings, space was given to papers on ethnological matters, even on occasion to physical anthropometry. But the struggle for scientific status by proponents of the human sciences before 1871 left geography as a section often devoid of content which took seriously the distributional study of humans as geographical subjects and, crucially, took humans seriously as geographical agents. Much study of humans after 1871 thus took place outside section E. Only later in the nineteenth century and in the early twentieth century did the geographical study of humans become, in contemporaries' usage, the study of human geography and the concern of that section.

Examples from the field: ethnographic survey and human geography, 1857–c. 1930

Guided by *Queries*, kindred publications such as *Hints to Travellers* and, from 1874, by the Association's *Notes and Queries on Anthropology, for the Use of Travellers and Residents in Uncivilized Lands* (which included Charles Darwin amongst its several compilers), those within the BAAS seeking for various reasons to measure humans looked to different locations in order to do so. Annual meetings provided talking space and an opportunity for expeditionary in-the-field measurement, particularly where easy access was afforded to 'the natives'. Examples from excursion science in Ireland, north-west Canada and South Africa illustrate these points.

For ethnologically-minded BAAS delegates in Dublin in 1857, their focus was less, perhaps, the papers on comparative Celtic craniometry and more the west of Ireland. They were not alone. When the German cartographer Gustav Kombst depicted the conjunction of linguistic, philological and ethnological distributions across Europe, he was, like others, posing questions about Europe's races, and, crucially, about the Celts (see Figure 6.1). What were the intrinsic features of what were held then to be archaic and barbarous languages spoken by peoples on Europe's western margins? Did philological and linguistic difference have racial correlates? Was the Celtic presence in Britain apparent in physical differences, in skull shape, for example, as well as in matters of culture? If in nineteenth-century Britain human science was racial science, its proving ground with respect to its theories and practices of measurement was Ireland, Wales and Highland Scotland: in fictitious historiographies of Anglo-Saxon supremacy, in anthropometric cartographies and in the deployment of racial 'scientific' indicators such as the 'Index of Nigresence' which typified the Celt as inferior in every way to the Anglo-Saxon English.[26]

The BAAS 'Ethnological Excursion' that went to Galway and the west of Ireland in the autumn of 1857 targeted the culture of the local people and, in accordance with anthropological interests, locals' physical dimensions.[27] The results were later co-ordinated by the Anthropological Laboratory in Trinity College, Dublin, which had opened in 1891, and it was from there that a series of related expeditions and surveys were undertaken of west Ireland's Celts in the 1890s. This work concentrated upon 'anthropography', the science of human description – looking first at hair and eye colour and physical measurements of the skull before turning to linguistic practices – and then to questions of sociology, folklore and ethnology (Figures 6.3–6.4). This work was under the direction of A. C. Haddon, Professor of Zoology at the Royal College of Science in Dublin.[28] Haddon's and others' work was part of a series of ethnographical surveys within Britain, the development of techniques for human measurement and of moves towards the establishment (never realised) of an Imperial Bureau of Ethnology for Greater Britain. For if the BAAS was concerned to document matters of race and cultural difference at home, it was doubly so of Britain's imperial subject peoples when it had the chance to meet overseas.[29]

That is why, and in keeping with practice, when 'the subject of Canadian anthropology came frequently under public and private discussion' as happened at the 1884 Montreal meeting, the decision was made to produce queries to direct the fieldwork. As with other ethnographic work, this was motivated by a sense of scientific urgency – expressed as early as 1839 by Prichard – lest cultural variety be lost in the face of civilisation:

> The opinion was strongly expressed that an effort should be made to record as perfectly as possible the characteristics and condition of the

Measurement, exploration and ethnology 179

Figure 6.3 BAAS ethnology in the field. The topmost image (No. 5) shows Michael Connelly, Inishmann, described by the ethnologists as 'A burly man, with the largest head measured in the Middle island'; No. 6 (centre) shows 'A characteristic group of the young men of Aranmore'; and No. 7 (at foot) Michael Flaherty and two women, Inishmaan. 'Flaherty refused to be measured, and the women would not even tell us their names'

Figure 6.4 Celtic physiognomy and 'racial types'. These images were among many taken during BAAS-associated ethnographic fieldwork in west Ireland in the 1890s

native tribes of the Dominion before their racial peculiarities became less distinguishable through intermarriage and dispersion, and before contact with civilized men has further obliterated the remains of their original arts, customs, and beliefs.[30]

The Canadian ethnological committee – which included E. B. Tylor, author in 1871 of the influential *Primitive Culture* – looked to the practical advantages that the Canadian Pacific Railway afforded in giving access to a number of native tribes 'whose languages and mode of life offer a field of enquiry as yet but imperfectly worked'. And as the British looked to the French and others, the BAAS additionally cited comparable work in the United States – where, under John Wesley Powell and others, 'the anthropology of the indigenous tribes has for years past been treated as a subject of national importance' – in order to engineer further government support.[31] Echoing Prichard's *Queries*, the memorandum of guidance issued by the Canadian committee 'has been drawn up for circulation amongst the Government officers in contact with the native tribes, medical practitioners, missionaries, colonists, and travellers likely to possess or obtain trustworthy information'. Given that the area to be covered was that 'vast region between Lake Huron and the Pacific',[32] it is not surprising that this committee had to be re-established, with more modest aims, in later BAAS meetings. Attention turned to the Canadian north-west, given the 'rapid disappearance of old customs, dress and mode of living' there in the face of European immigration and because, in 1899 anyway, the BAAS could rely upon 'the present availability of the services of an expert and enthusiastic observer'[33] (a Mr C. Hill-Tout).

Comparable work for South Africa was partly the concern of section H, but, as the overseas meetings also provided a means for the assessment of the geographical potential of the dominions, the empire's peoples there were also studied by the geographers in section E. One expression of this was the Committee on the Human Geography of Tropical Africa, which operated between 1926 and 1939 under the chairmanship of Sir Charles Lucas. The committee's brief was 'To enquire into the present state of geographical knowledge of Tropical Africa and to make recommendations for its furtherance and development'.[34] Although initially concerned with the absence of proper maps to facilitate such development – thus exemplifying that sense, noted above, that maps and mapping provided the ground work against which other forms of geographical science could be conducted – the committee focused mainly on securing reliable knowledge about Africa in order to shape imperial policy. Recognising that much material on African peoples had been collected by ethnologists and anthropologists, the committee noted an important missing element, 'namely, the treatment of matter which may be classed as *social geography*':

> The life of native groups as a whole is incomprehensible unless it be studied in relation to the particular kind of land in which they live, and it is most unlikely that adequate attention will be devoted to this intimate relationship between man and his environment in British Tropical territories unless British geographers endeavour to see that much greater stress is laid upon this geographical question than has hitherto been the case.[35]

The committee hoped to realise its aims by ensuring work was done 'in Great Britain or in Africa by members of the Committee and their helpers' and 'by inducing residents in Africa to undertake local investigations of an anthropo-geographical character'. For Lucas, 'The most urgent need of the Committee is, therefore, for some method of explaining to residents in Africa the nature of the information sought.' To this end, the committee proposed to circulate copies of 'one or two of the very few brief studies that have hitherto been made upon the human geography of African regions' – they cited French geographical work rather than earlier ethnological guides – and by issuing a questionnaire 'which the local contributors would be asked to answer . . . with the request that as much information as possible be furnished in map form'.[36] Pamphlets were produced, copies being sent out '(1) In bulk to the Governments of the various British Colonies and Mandates in Tropical Africa and . . . distributed by them to their Local Officers, and (2) To a selected number of missionaries and other residents'.[37]

There is an unbroken methodological thread between the 1839 *Queries* and the pamphlet issued nearly a century later by this subcommittee of section E. The human sciences as a whole were shaped by these instructional directives and, as with their physical science counterparts, by the epistemological reliance placed upon the observational skills and instrumental capacities of appropriate people. When, in the years after about 1890 and more distinctly in the early twentieth century, a more evidently human geography as a whole and more particular varieties of human geography started to emerge (Table 3.3), it and they did not do so simply because of developments within section E and from the development of procedural practices and sectional identities shaped by decades of dispute over how best to study and measure humans. To understand the shift from the geography of humans to human geography we need to look beyond the BAAS and the activities of section E.

The becoming of (human) geography, c. 1885–1933

The history and geography of geography's institutionalisation and formalisation as a university subject within Britain in and from the later nineteenth century has been told elsewhere and it is not my intention to repeat it here.[38] But I do want to claim that what was taking place within the BAAS after about 1885 in terms of the development of human geography was directly related to the foundation of teaching departments for the subject as a whole and to the related emergence of a body of professional geographers who, by virtue of their educational training and the establishment of research agenda, began to teach, practise and promote more specialist forms of geography.

Key elements in this heightened profile for geography and human geography were the Keltie Report of 1886, Halford Mackinder's 1887 paper on the scope and methods of geography as both a descriptive and a synthesising science, the establishment of teaching departments for the subject at Oxford, in 1887, and at Cambridge in 1888, and the foundation of departments of geography throughout Britain from the early years of the twentieth century: in Manchester in 1892 (then, only as a lectureship), in Aberystwyth in 1906, Reading (1907) and in Sheffield and in Edinburgh (1908), Leeds, Liverpool and Glasgow (1909), Southampton (1913), the London School of Social and Political Science (1918) and Aberdeen (1919). Even by 1920, however, there were only six professorships in geography: at University College, London (1903), Reading (1907), Oxford (1910), Aberystwyth and Liverpool (in 1917) and at Birkbeck College, London (1920). If, for one leading scholar of the history of British geography, the recency of many of these institutional developments is 'salutary' when considering geography's subsequent place in British education,[39] understanding their role in the historical present is important in explaining the nature and place of geography within the BAAS.

What has often been seen as initiatives undertaken only by the Royal Geographical Society in promoting this 'New Geography' from the mid 1880s was at least in part an agenda shared by the BAAS. The important link between the two bodies was John Scott Keltie. Keltie's role as an educationalist and advocate of geography is acknowledged in relation to the RGS (to which he was secretary for over thirty years) and of the Geographical Association, founded in 1895 to promote geography in schools (he was president of that body in 1913). But Keltie was also secretary to section E between 1885 and 1892 (and president in 1897 and section vice-president ten times). The BAAS secretaryship, like that of the RGS, was an influential position, given developments over the fortunes of geography as an educational subject from the 1880s. Keltie's

1886 report, commissioned in 1884, noted the relatively poor state of geographical education in British schools, particularly in comparison with geography's place in Germany and in France, and presented a convincing case for geography as a subject that was important to teach in order that an imperial nation and citizenry might know its own bounds and those of others. Further, it promoted the view that geography was a systematic science, capable of being examined and distinguished by principles of distribution, causal explanation and with regional variability and relationships between humans and the environment as its distinguishing focus.[40]

At its general business meeting in January 1887 the BAAS endorsed the findings of a sub-committee 'appointed for the purpose of cooperating with the Royal Geographical Society in endeavouring to bring before the authorities of the Universities of Oxford and Cambridge the advisability of promoting the study of Geography by establishing special Chairs for the purpose'. In letters to the vice-chancellors at Oxford and at Cambridge, Lord Aberdare, BAAS president, and Richard Strachey, BAAS vice-president (both also leading figures in the RGS) stressed the importance of geography and its political and broader intellectual utility:

> So much of human knowledge and human interests is bound up with the relations and interaction of the physical conditions of the earth, the study of which is practically embraced in geography, that there are few branches of education which do not present a geographical aspect, and which do not therefore offer a field for instruction in geography in combination with some other subject.
>
> It is unnecessary to insist upon the close connection of history and geography, or upon the importance of a knowledge of the physical conditions of the various regions of the world, to those who engage in the conduct of our political affairs.
>
> Without the comprehensive study of the earth, for which Englishmen [sic], as a people, have the largest opportunities and the least preparation, physical students would fail to grasp the true character and relations of the various sciences of observation, such as anthropology, geology, botany, meteorology, &c.[41]

Keltie, the moving force behind this initiative, had made similar representation via the RGS in 1871 and again in 1874. Keltie made these representations between 1871 and 1886 almost wholly through the RGS and not via the BAAS or via section E: within the BAAS, he spoke on 'Geographical Education' only to the 1885 Aberdeen meeting, drawing upon his own report and, in part, establishing his authority from his position as 'Inspector of Geographical Education', an appointment made by the RGS in 1884 to lend credence to its support of Keltie's research on geography, education and science. In the BAAS between 1871 and 1885, aside from Keltie's Aberdeen paper, geography's place in education was not the subject of detailed papers or of a recommendation, nor was it even regularly on the section committee's agenda. What more evidently occupied section E in its meetings were papers on exploration, cartography and topographic mapping (notably of the Near East) and, in association with the newly formed section H, papers on anthropology. These papers in particular, to judge from newspaper reports, were much better attended and enjoyed than their geographical counterparts: E. B. Tylor's 'very able lecture' in Edinburgh in August 1871 on the relation of primitive and 'primeval man' to modern civilisation was the occasion of 'considerable amusement'; another paper that year on 'Man and the Ape' was interpreted differently by different members of the audience in relation to Darwin's recently published *The Descent of Man* (1871); and we should recall that it was H. M. Stanley and not Francis Galton who drew the crowds at Brighton in 1872. There was nothing about Keltie's report in section E sessions other than in 1885: in short, what filled the lecture programmes and the halls was anthropology, mapping and exploration, not discussions about the state of geography in British education.[42] Keltie's BAAS work was strategic, not a matter of public acclaim or, between 1871 and 1885 at least, even of public debate.

Following the establishment of teaching programmes at Oxford and at Cambridge, the development of a formal geographical curriculum remained uneven. At the 1895 Ipswich BAAS meeting, delegates to the geography sessions did debate the still uncertain place of geography in the educational system, partly to review the effect of the subject's gathering presence within these universities since the later 1880s and partly in response to Sir Clements Markham, who, at the Sixth International Geographical Congress that year in London, had spoken 'most impressively of the inadequate manner in which geography was treated in our country, and urged the need of altering this'. The BAAS sub-committee appointed to review the 'Position of Geography in the Educational System of the Country' (a group chaired by Mackinder, with A. J. Herbertson as its secretary, and with H. R. Mill, J. S. Keltie, E. G. Ravenstein and E. Sowerbutts as its other members) recognised that there was more work to do yet but that it, and the BAAS, were hindered by lack of funds: 'the Committee, owing to there being no funds at their disposal, have not been able to undertake a personal inspection of various educational institutions at home and abroad, such as that carried out by Dr. Keltie'.[43]

Two years later, however, as he gave the section E presidential address in Toronto, there was – in Keltie's view at least – a strong sense that geography's intellectual and educational status had been secured and that the causes and the consequences of this were clear:

> Better methods have been introduced in our schools; a much wider scope has been given to the subject; in many quarters teachers have shown themselves anxious to be guided in the right direction; and, above all, both Oxford and Cambridge at length consented to the establishment of lectureships in geography. A school of young geographers has grown up, consisting of men who have had a thorough university training in science and letters, and who are devoting themselves to the various branches of geography as a specialty. In this way the arid old text-books and character-less maps are being supplanted by others that will bear comparison with best productions of Germany. Photography and lantern slides illustrating special geographical features are coming into use in schools; and in other directions appliances for use in education are being multiplied and improved. A British geographical literature is growing up, and if, as I hope, the progress be maintained, we shall be able to hold our own in geography with any country. The interest in the subject has been extended by the foundation of geographical societies in various large centres; whereas thirteen years ago the only geographical society was that of London, there are now similar societies in Manchester, Newcastle, Liverpool, and Edinburgh, the last with branches in Glasgow, Dundee, and Aberdeen. If this progressive movement is maintained, as there is much reason to hope it will be, the scientific and educational aspects of geography in Britain will be more nearly on a par with exploration in which our country has so long held the lead.[44]

Keltie's claims and aspirations are, broadly, borne out by what we know from studies of the place of geography in British school and university education, of aspects of geography's educational technology, and on the role of the Royal Scottish Geographical Society, established in 1884, and of the several provincial geographical societies in England.[45]

The evidence of BAAS geography grants awarded and of papers presented (Tables 3.2–3) allows us to assess his claims in more detail. With respect to the papers presented, there was, as we have noted, an uneven specialisation notably with respect to human geography after 1911, more clearly so in the period 1921–33: in regional geography, in social geography and, less strongly, in historical geography (Table 3.3). Papers on commercial geography and on urban geography were also more commonly given after about 1911 than had earlier been the case. Here, in BAAS meetings after about 1911, is evidence that Keltie's 'school of young geographers' was at work, presenting an increasingly specialised and thematic as well as an emergent regional focus to their work. Yet we must be cautious about seeing this as a wholesale shift away from generalist papers and those of an exploratory focus towards 'modernity' and disciplinary specialisation – not least since evidence discussed in the following chapter complicates the picture yet further with regards to geography's scientific status and popular interest in the topic in the period after 1911. There is, nevertheless, a discernible increase in the proportion of papers on geography's human specialisations, on its place in education, its connections with other sciences and the proponents, in the main, were men formally employed as geographers.

The professionalisation of geography's personnel engaging with the BAAS is evident from the grants awarded (Table 3.2). After about 1910, many of the successful applicants for funding in human and in physical geography were established in university positions: H. O. Beckit (Oxford), Charles Fawcett (lecturer at Southampton 1913–19, then at Leeds 1919–1928), J. McFarlane (Aberdeen), Percy Maude Roxby (founder of the Liverpool department in 1909), Frank Debenham (Cambridge), George Goudie Chisholm (first lecturer at Edinburgh, a specialist in 'commercial geography', forerunner to economic geography even as he was funded by the BAAS for cartographic survey work in Spitsbergen) and Robert Rudmose-Brown, Antarctic explorer-become-professional geographer, first at Manchester between 1920, then as professor at Sheffield from 1931. For Stanley Beaver, the transition in the nature of BAAS section E was evident in and from 1913: 'a landmark [meeting] in the Section's history' in that it had an academic president (H. N. Dickson) and 'several papers by young geographers whose names subsequently became household words in academic circles' (he cites A. G. Ogilvie, C. B. Fawcett, Hilda Rodwell Jones and A. J. Herbertson amongst others) (Figure 6.5).

But geography's changed presence in the BAAS was most marked after 1919 and before 1939: 'This was the period when geography was establishing itself as an Honours school in many universities, and Presidential addresses were often wide-sweeping reviews of the discipline and its functions':

> Thus Marion Newbigin, in 1922, spoke on 'Human geography: first principles and some applications'; Roxby in 1930 on 'the scope and aims of human geography'; Mackinder in 1931 on 'The human habitat'; Fleure in 1932 on 'The geographical study of society and world problems'; Fawcett in 1937 on 'the changing distribution of population'; and Griffith Taylor in a characteristically individualistic address in 1938 on 'Correlation and culture'.

Yet – and importantly in relation to that other evidence presented here – 'No Presidential Address during this period took a physical theme, though there were two on cartographical topics, and both Rudmose

Figure 6.5 Portrait of Marion Newbigin, BAAS section E president, 1922

Brown and Debenham chose the Polar regions as their topics.'[46] In short, although it is the case that specialist work within section E began from the later 1890s and early 1900s to fill the gaps as papers on and grants for exploration waned, human geography within the BAAS developed after about 1910 and more in consequence of what was happening elsewhere – in the professionalisation, institutionalisation and formalisation of geography as a subject for teaching and research – than it did from those longer-running discussions of the place of geography, ethnology and the anthropology of humans within section E and section H.

Conclusion

This chapter has demonstrated with reference to geography and the human sciences how the BAAS worked from the mid 1830s to establish its own scientific status and, within that institutional narrative, how different sciences and practices sought to establish themselves. The chronology of geography's sectional history – within section C from 1836 until 1851, jointly with ethnology between 1851 and 1871 and as a section in its own right thereafter – has been shown to be a 'messy' story about the range and nature of geography as a subject concerned with the measurement of humans and human culture.

In respect of its place in the human sciences, what was held to be geography before and after 1851 was associated to varying degrees with ethnographic description and with more strictly anthropological even anthropometric measurement work, and took much of its human content from explorers' narratives and expeditionary accounts. In contrast to the largely uninhabited polar regions, the study of humans from within the BAAS turned to the tropics, to 'exotic' colonial peoples and to peoples nearer to home whose culture, language, hair colour and skull shape marked them as different. Herein lay one reason for geography's enduring popularity and, in the eyes of many within the BAAS and associated institutions, also of geography's practical utility. To the Victorian imagination, the far-away fascinated; far-away peoples and domestic oddities nearer to home fascinated even more.[47] As Murchison influenced the shape of the section and the content of the science, so the development of guides to ethnographic study and the photographic portraiture of human difference linked emergent geographers with ethnologists, anthropologists and social statisticians and so, too, the languages and practices of the human sciences became part of the language and practice of science as a whole.

What shaped 'human geography' in this period, certainly before mid century and particularly in respect of the study of human sciences, was less any intrinsic human content to geography and rather more those related debates over human culture and human physical differences which helped mould ethnology and anthropology, notably from 1871. Of course, section E incorporated elements of ethnological work before as well as long after 1871. Similarly, section H – known by some in the 1870s as the 'Ape Section', given its excessive attention to the comparative anatomy of apes and men, and, often, the racialised physiognomy of colonial natives as 'beneath and below' their imperial masters – included papers in which geography was understood as the study of spatial distributions.[48] As we have seen, when the controversial Paul Du Chaillu

Figure 6.6 'My first gorilla.' Paul du Chaillu made much of his African explorations in his BAAS and other public talks. The gorilla looks startled, or is about to break into song

talked on his exploits with gorillas and tribesmen in equatorial Africa to the Manchester meeting in 1861, he did so to large crowds, to different sections (E and D) and to decidedly mixed reactions. To the backdrop of a mock interview between the explorer and a gorilla, and maps of Africa, the public egged him on 'with his colourful story': others, ethnologists and anthropologists alike, were more sceptical of his claims, given the many discrepancies in his account[49] (Figure 6.6). Given the intrinsically difficult questions of seeing things for themselves, geographers and others could be hoodwinked by charlatans peddling popular science: the lecture on Australian peoples and aboriginal customs by Louis de Rougement (real name Henri Louis Grin, a long-time hoaxer) to the section E proceedings at Bristol in 1898 was afterwards exposed in the press as largely fraudulent (Figure 6.7). Its popularity at the time was paralleled from the start by scepticism from amongst geography's professionals, at least one of whom thought de Rougement 'a mendacious rascal'.[50]

The human sciences were shaped by competing claims to the place and diversity of man and were so in particular BAAS meetings: in Birmingham in 1839 and 1865, for example, and in Bath in 1864. Consider, too, the debates in Nottingham in August 1866 following Reddie's paper 'On the Serious Theories of Man's Past and Present Condition'. This was a lengthy address which offered comparative

Measurement, exploration and ethnology 191

Figure 6.7 Louis de Rougemont, geographical hoaxer. De Rougemont's talk to section E at the 1898 Bristol BAAS meeting occasioned great interest among the public present and scepticism among the professional geographers

scrutiny of religious, Darwinian and polygenist theories of human origins. Thomas Huxley dismissed the paper as 'purely anthropological' and so reckoned that it ought not to be in section E. For Crawfurd, anthropology and ethnology were synonymous 'or nearly so': his criticism was targeted at the Darwinian element of the speech, since he 'doesn't believe any of Darwin's theory' (a remark that occasioned numerous 'hear, hear's from the audience).[51] Such moments were part of a more complex and varying context in which the study of humans was central to the make-up, and certainly to the popularity, of the BAAS, but one in which discussions of the geography of humans did not always take place within the confines of section E.

The more evidently human (or 'social') geography that began to find an expression within BAAS section E from the last years of the nineteenth century and early years of the twentieth (*cf.* Table 3.3) did so in part because it built upon others' study of the geography of humans and because it shared methodological procedures with the other human sciences. Most crucially, these features reflected developments after about 1845 over the institutional and professional status of geography and the appearance in greater numbers than ever before of professional geographers in the stead of 'professed geographers'. Human geography emerged too because of a recognition that humans could be seriously studied in different geographical settings and because, by then, the study of man was less concerned with humans *per se* in their physiognomic or ethnological differences and more with human agency and with human–environment interaction. One expression of these concerns was, as we have seen, Lucas's attention in 1914 to man as a geographical agent. Another was the slow emergence in geography of regional synthesis, the science that offered a bridge between the human and the natural environments beyond the facts of distributed measurements alone. Speaking in 1922 about human geography's 'first principles' as she called them, Marion Newbigin considered just such matters:

> It seems to me, therefore, that the most clamant need at the present time is a continuous attempt to make it plain to the community at large that the main interest of geography is not in its facts as such – for if geography ceased to exist the geologists, meteorologists, botanists, zoologists, and so forth would continue to collect most of these. Rather does it lie in the way in which the geographer studies these facts in their relations to each other and to the life of man.[52]

As the next chapter shows, Newbigin's appeal was made in a context that, by then, placed great value upon geography as a civilising science, as a means to citizenship. Yet even as geography offered the promise of political and intellectual utility in being in early twentieth-century Britain and in the dominions of empire a discipline of synthesis, countervailing tendencies were at work that cast doubts over its credentials as a science at all.

Notes

1 In a voluminous literature I have found the following helpful in this context: R. Yeo, *Defining Science: William Whewell, Natural Knowledge and Public Debate in Victorian Britain* (Cambridge: Cambridge University Press, 1993); R. Yeo, *Science in the Public Sphere: Natural Knowledge in British Culture, 1800–1860* (Aldershot and Burlington VT: Ashgate, 2001); M. Daunton (ed.), *The Organisation of Knowledge in Victorian Britain* (Oxford: Oxford University Press in association with the British Academy); J. Golinski, *Making Natural Knowledge: Constructivism and the History of Science* (Cambridge: Cambridge University Press, 1998); T. M. Porter and D. Ross (eds), *The Cambridge History of Science*, Vol. 7, *The Modern Social Sciences* (Cambridge: Cambridge University Press, 2003); D. N. Livingstone, *Adam's Ancestors: Race, Religion and the Politics of Human Origins* (Baltimore MD: Johns Hopkins University Press, 2008); A. Desmond and J. Moore, *Darwin's Sacred Cause: Race, Slavery and the Quest for Human Origins* (London: Allen Lane, 2009); G. Stocking, *Victorian Anthropology* (New York: Free Press, 1987); M. J. S. Rudwick, *Bursting the Limits of Time* (Chicago: University of Chicago Press, 2005); M. J. S. Rudwick, *Worlds before Adam* (Chicago: University of Chicago Press, 2008).
2 On this point, see L. Daston and P. Galison, *Objectivity* (New York: Zone Books, 2007).
3 Desmond and Moore, *Darwin's Sacred Cause*, pp. 144–5.
4 [A committee of the British Association for the Advancement of Science], *Queries respecting the Human Race, to be addressed to Travellers and Others* (London: Richard & John Taylor, 1839).
5 On the birth of the social fact in these terms, see M. Poovey, *A History of the Modern Fact: Problems of Knowledge in the Sciences of Wealth and Society* (Chicago: University of Chicago Press, 1998); T. Porter, *Trust in Numbers: The Pursuit of Objectivity in Science and Public Life* (Princeton NJ: Princeton University Press, 1995); A. Rusnock, *Vital Accounts: Quantifying Health and Population in Eighteenth-Century England and France* (Cambridge: Cambridge University Press, 2002). For a discussion of *Hints to Travellers* within the history of regulating observation and fieldwork in geography, see Driver, *Geography Militant*, pp. 40–1, 49–50, 56–67.
6 M. Friendly and G. Palsky, 'Visualising nature and society', in J. R. Akerman and R. W. Karrow (eds), *Maps: Finding our Place in the World* (Chicago: University of Chicago Press, 2007), pp. 207–53.
7 Livingstone, *Adam's Ancestors*, pp. 109–36; A. Kuper, 'Anthropology', in

Porter and Ross (eds), *Cambridge History of Science*, Vol. 7, pp. 354–78; Stocking, *Victorian Anthropology*.
8 Desmond and Moore, *Darwin's Sacred Cause*, p. 154.
9 Royal Anthropological Institute, A1, Minutes of the Ethnological Society (2 January 1844–26 January 1869).
10 O. J. R. Howarth, 'The centenary of Section E (Geography)', *Advancement of Science* 8 (1951), pp. 151–65, quote at p. 152.
11 Stafford, *Scientist of Empire*, pp. 20, 187, 221.
12 Royal Geographical Society Archives, Correspondence Block CB 1841–1850, R. Murchison, 10 August 1850.
13 R. I. Murchison, 'Geography and ethnology', *Report of the Twenty-eighth Meeting of the British Association for the Advancement of Science* (London: John Murray, 1859), pp. 143–4, quote at p. 144.
14 J. Crawfurd, 'On the connexion between ethnology and physical geography', *Report of the Thirty-first Meeting of the British Association for the Advancement of Science* (London: John Murray, 1862), pp. 177–83, quote at p. 177; J. Crawfurd, 'On the connexion between ethnology and physical geography', *Transactions of the Ethnological Society of London* 2 (1863), pp. 4–23.
15 Livingstone, *Adams' Ancestors*, pp. 112–14; Stocking, *Victorian Anthropology*, pp. 250–1.
16 R. I. Murchison, 'Geography and ethnology', *Report of the Thirty-third Meeting of the British Association for the Advancement of Science* (London: John Murray, 1864), pp. 126–33, quote at pp. 132–3. On the later printing of Crawfurd's 1861 address to section E, see J. Crawfurd, 'On the connexion between ethnology and physical geography', *Transactions of the Ethnological Society of London* 2 (1863), pp. 4–23.
17 C. Nicholson, 'Geography and ethnology', *Report of the Thirty-sixth Meeting of the British Association for the Advancement of Science* (London: John Murray, 1867), pp. 98–9, quote at p. 98.
18 S. Baker, 'Geography and ethnology', *Report of the Thirty-seventh Meeting of the British Association for the Advancement of Science* (London: John Murray, 1868), pp. 104–11, quote at p. 105.
19 Royal Anthropological Institute, A3:-1, Minutes of the Anthropological Society (8 January 1863–18 January 1870), A8/2/2, and see also J. Hunt, 'The President's Address', *Journal of the Anthropological Society of London* 3 (1865), pp. lxxxv–cxiii, quote at p. lxxxvii.
20 Royal Anthropological Institute, A3:-1, Minutes of the Anthropological Society (8 January 1863–18 January 1870), and, in particular, A8/2/1, Letter of James Hunt to BAAS, 22 August 1865, and A8/2/2, Address to the Anthropological Society of London, 3 January 1865.
21 [Anon.], 'Anthropology and the British Association', *Anthropological Review* 3 (1865), pp. 354–71.
22 Royal Geographical Society, RGS SSC/11 File 2, Minute Book of Section E (Geography and Ethnology), 1865–1873, Saturday 9 September 1865.

23 A. Turner, 'Anthropology', *Report of the Forty-first Meeting of the British Association for the Advancement of Science* (London: John Murray, 1872), pp. 145–7, quote at p. 145.
24 For a contemporary account of these issues, see [Anon.], 'Anthropology and the British Association', *Anthropological Review* 3 (1865), pp. 354–71.
25 R. I. Murchison, 'Geography', *Report of the Fortieth Meeting of the British Association for the Advancement of Science* (London: John Murray, 1871), pp. 158–66, quote at p. 158.
26 J. Urry, 'Englishmen, Celts and Iberians: the Ethnographic Survey of the United Kingdom, 1892–1899', in G. Stocking (ed.), *Functionalism Historicised: Essays on British Social Anthropology* (Madison WI: University of Wisconsin Press, 1984), pp. 83–105; P. Gruffudd, 'Back to the land: historiography, rurality and the nation in interwar Wales', *Transactions of the Institute of British Geographers* 19 (1994), pp. 61–77; C. W. J. Withers, *Geography, Science and National Identity: Scotland since 1520* (Cambridge: Cambridge University Press, 2001), pp. 220–5; H. Winlow, 'Anthropometric cartography: constructing Scottish racial identity in the early twentieth century', *Journal of Historical Geography* 27 (2001), pp. 507–28.
27 [Anon.], *A Short Description of the Western Islands of Aran, County of Galway, chiefly extracted from the Programme of the Ethnological Excursion of the British Association to these interesting Islands in the Autumn of 1857, under the direction of W. R. Wilde, M.R.I.A.* (Dublin: no publisher, 1858); M. Haverty, *The Aran Isles; or, A Report of the Excursion of the Ethnological Section of the British Association from Dublin to the Western Islands of Aran in September, 1857* (Dublin: no publisher, 1859).
28 A. C. Haddon and C. R. Browne, 'On the ethnography of the Aran islands, County Galway', *Proceedings of the Royal Irish Academy* 2 (1892), pp. 768–830. This was the first of seven such reports between 1892 and 1902.
29 Urry, 'Englishmen, Celts and Iberians'.
30 [British Association for the Advancement of Science], *Committee on North-Western Tribes for the Dominion of Canada* (Montreal: British Association for the Advancement of Science, 1884), p. 1.
31 *Committee on North-Western Tribes*, p. 1; S. Kirsch, 'John Wesley Powell and the mapping of the Colorado plateau, 1869–1879: survey science, geographical solutions, and the economy of environmental values', *Annals of the Association of American Geographers* 92 (2002), pp. 548–72; D. Worster, *A River Running West: The Life of John Wesley Powell* (Oxford: Oxford University Press, 2001), pp. 262–94.
32 *Committee on North-Western Tribes*, p. 2.
33 [British Association for the Advancement of Science], *Ethnological Survey of Canada* (London: John Murray, 1899), pp. 1–2.
34 MS. Dep. BAAS 333, fol. 41. The committee members were Sir Charles Lucas (chair), Mr Alan Ogilvie (secretary), Mr W. H. Barker, Professor P. M. Roxby and Mr John McFarlane.

35 MS. Dep. BAAS 333, fol. 72 [Report of the Committee on the Human Geography of Tropical Africa, Johannesburg 1929].
36 MS. Dep. BAAS 333, fols 76–7.
37 MS. Dep. BAAS 333, fol. 158. The questionnaire was separately published as *The Human Geography of Tropical Africa: The Need for Investigation* (London: British Association for the Advancement of Science, 1930): see MS. Dep. BAAS 333, fols 95–8.
38 For summary accounts in a wide range of works on the topic, see D. Livingstone, 'British geography, 1500–1900: an imprecise review', in R. Johnston and M. Williams (eds), *A Century of British Geography* (Oxford: British Academy in association with Oxford University Press, 2003), pp. 11–44; R. Johnston, 'The institutionalisation of geography as an academic discipline', in Johnston and Williams (eds), *A Century of British Geography*, pp. 45–90; C. W. J. Withers, 'A partial biography: the formalization and institutionalization of geography in Britain since 1887', in G. S. Dunbar (ed.), *Geography: Discipline, Profession and Subject since 1870* (Dordrecht: Kluwer, 2001), pp. 79–119.
39 D. R. Stoddart, *On Geography* (Oxford: Blackwell, 1986), p. 46. On Mackinder's influential 1887 paper, see H. J. Mackinder, 'On the scope and methods of geography', *Proceedings of the Royal Geographical Society* 9 (1887), pp. 141–60.
40 J. S. Keltie, 'Geographical education: report to the Council of the Royal Geographical Society', *Supplementary Paper of the Royal Geographical Society* 1 (1886), pp. 439–594.
41 [Lord Aberdare], 'On the promotion of the study of geography', *Report of the Fifty-seventh Meeting of the British Association for the Advancement of Science* (London: John Murray, 1888), pp. 158–9.
42 For reports on Tylor's 1871 Edinburgh lecture, see *The Daily Review*, 8 August 1871, p. 2; *The Edinburgh Evening Courant* described the anthropological lectures as 'crowded to suffocation' (4 August 1871, p. 3).
43 H. J. Mackinder *et. al.*, 'The position of geography in the educational system of the country', *Report of the Sixty-seventh Meeting of the British Association for the Advancement of Science* (London: John Murray, 1898), pp. 370–409, quote at p. 370.
44 J. S. Keltie, 'Geography', *Report of the Sixty-seventh Meeting of the British Association for the Advancement of Science* (London: John Murray, 1898), pp. 699–712, quote at p. 701.
45 See, for example, R. Walford, *Geography in British Schools, 1850–2000* (London: Woburn Press, 2001); T. Ploszajska, *Geographical Education, Empire and Citizenship: Geographical Teaching and Learning in English Schools, 1870–1944* (Cambridge: Historical Geography Research Group Publication Series 35, 1999); Withers, 'A partial biography'; Johnston, 'The institutionalization of geography as an academic discipline'; E. Lochhead, 'The Royal Scottish Geographical Society: the setting and sources of its success', *Scottish Geographical Magazine* 113 (1997), pp.

42–50; Withers, *Geography, Science and National Identity*, pp. 195–235; T. N. L. Brown, *The History of the Manchester Geographical Society, 1884–1950* (Manchester: Manchester University Press, 1971); J. M. MacKenzie, 'The provincial geographical societies in Britain, 1884–1914', in M. Bell, R. A. Butlin and M. J. Heffernan (eds), *Geography and Imperialism, 1820–1940* (Manchester; Manchester University Press, 1995), pp. 93–124.
46 The quotes are from S. Beaver, 'Geography in the British Association for the Advancement of Science', *Geographical Journal* 148 (1982), pp. 175, 176.
47 On this, see for example J. Schwartz, *'The Geography Lesson:* photographs and the construction of imaginative geographies', *Journal of Historical Geography* 22 (1996), pp. 16–45; J. R. Ryan, *Picturing Empire: Photography and the Visualization of the British Empire* (London: Reaktion, 1997).
48 The designation 'the Ape Department' is given of section H at the 1879 Sheffield meeting: *Sheffield Daily Telegraph*, 22 August 1879, p. 4.
49 See, for example, reports in the *Manchester Courier and Lancashire General Advertiser*, 7 September 1861.
50 H. R. Mill, *Life Interests of a Geographer, 1861–1944: An Experiment in Autobiography* (East Grinstead: published privately, 1944), p. 99.
51 *Nottingham Daily Guardian*, 27 August 1866.
52 M. Newbigin, 'Human geography: first principles and some applications', *Report of the Ninety-second Meeting of the British Association for the Advancement of Science* (London: John Murray, 1922), pp. 94–105, quote at p. 95.

7

Science, education and the 'crisis' in geography, 1910–c. 1939

Discussing the British Association's later history, MacLeod has observed that the organisation went through a 'period of indecision' from about 1905 to 1919, following a 'period of imperial colonisation in ideas and images' between the mid 1880s and about 1905. Between 1919 and 1930, the BAAS was marked by a 'period of retrospection': 'surrounded by specialist bodies which required no assistance in establishing professional credentials, but who welcomed greater public support, the Association helped transform the public impression of science, and was itself transformed by the press into a platform for popularising scientific knowledge'. Noting, too, how the years between 1931 and 1939 were characterised by intensified economic and social debate and public disenchantment with science and technology – circumstances characteristic of what others have termed 'the Morbid Age' in Britain – MacLeod remarked too how 'Some of these historical periods and issues have received close attention, others, far less.'[1]

Such claims are open to examination not only in terms of the dates proposed and their distinguishing criteria but also for different sciences within the organisation. This chapter extends from foregoing arguments here and from what little work has been done elsewhere on the Association in the early twentieth century to examine for geography MacLeod's thesis and his periodisation of its activities.[2] What we have already seen of geography's nature and status – matters of different practice and sectional identity, emergent professionalisation within and outwith the BAAS, connections between section E and the RGS and other bodies, moves to more thematic work, the attention paid to regional geography and the persistence of different views about geography's popularity and exploratory traditions, its 'scientific' content, and its role in relation to imperial concerns – might allow us, broadly, to concur with MacLeod's points about institutional specialisation, scientific credentialism and, albeit over a longer period, geography's popularism. But closer scrutiny of geography's nature and status as a science in the first third or so of the twentieth century reveals rather different chronologies and key moments by which its fortunes and its content as a BAAS subject were shaped in the early twentieth century.

The empirical focus of this chapter – and of the book – comes to an effective end with the foundation, in January 1933, of the Institute of British Geographers. The establishment of this research-led geographical organisation was, in part, a move born of frustration at the torpor of the RGS over the place of research and research-based publication and the identity of professional geographers within British geographical and scientific culture. In part, too, the creation of the IBG was a reaction against section E of the BAAS, but this was less strongly felt than it was for the RGS. At its first council meeting in January 1933 the new body agreed that nothing would be done which would conflict with the interests and purpose of section E, and for several of its early years the IBG met during BAAS annual meetings, or held field trips after that meeting: in its pattern of movement throughout the country, its shared membership and, latterly, in its institution of a presidential address, the IBG mirrored several of the features of BAAS section E.[3]

The establishment of the IBG thus signalled something of a 'new beginning' in geography's professional identity and associational credentials in Britain. But its importance here rests not in its foundation *per se* or in relation to its history after 1933 but, rather, as a convenient *terminus ad quem* against which to understand the nature of geography within section E in the inter-war period. Debates within section E and beyond in this period again addressed geography's purpose and focus, its characteristics as a science, and as a science of empire. But as geography's professional practitioners grew in number from the early twentieth century there is evidence to suggest that not all saw section E as an appropriate outlet for their labours. As we shall also see – and have already observed in relation to BAAS overseas meetings between 1884 and 1929 – geography in the BAAS in the 1920s and 1930s was caught up in wider debates about citizenship, internationalism and empire and was so by virtue of its popularity and its educational potential even as it was becoming more specialised, its purpose oriented more to research, education and publication than to popular acclaim and spoken tales of exploration. Rather than 1905 and 1919 as meaningful dates in the association's longer institutional history, for geography the years 1910, 1911 and 1912 mark key moments in the subject's modern scientific history. In the first, geography and its practitioners became embroiled in BAAS plans to restructure its scientific sections. The second identifies a moment of 'geographical heresy' when, and from within section E, geography was accused of not being scientific enough – even that it was hardly a science at all. The third is distinguished by differences over professional status and the local nature of BAAS meetings.

Scientific identity, 'geographical heresy' and geographies of local science: the events of 1910–1912

London 1910: sectional restructuring and geography's identity
The persistent distinction but not sharp disjuncture between geography's exploratory and scientific 'traditions' revealed throughout this study was by early 1910 again a matter of moment for section E officers. That year, the BAAS considered a sectional reorganisation that would have reunited geography and geology. Sectional reorganisation was in general terms predicated upon the need to make BAAS science more engaging to the public and to make management of sections easier, where consonant intellectual concerns would permit. More particularly, it was felt that 'one of the principal objects of the Association was to give publicity to papers on border line subjects, rather than those on subjects which might be dealt with by the society for any one definite science' and 'That more attention should be paid to the previous selection of subjects, with particular attention paid to the interests of the places of meeting; and that discussions should be more carefully arranged.' The move to amalgamate sections where it was sensible to do so was in part managerial: 'such groups might be used to lighten the work of the Committee of Recommendations by having a lump sum for grants allocated to each group'.[4]

Although it is hard to know whether such proposed reorganisation should be seen, to use MacLeod's terms, as born of indecision within the London headquarters or as restructuring with a view to better public provincial outreach, one agreed imperative was to ensure that the science in the sectional meetings was more closely attuned to the interests and demands of the places of the meeting. BAAS officers recognised that the development of subject-specific associations and of university departments meant that the BAAS had to change: 'The Science Committees of the B.A. have diminished in importance with the development of organized teaching in science at the Universities.'[5] For the BAAS as a whole, reorganisation was directed at providing a forum for public science. For subject officers within the BAAS whose sections were in line for amalgamation based more upon administrative efficiency than intellectual consonance, the reorganisation was not welcome.

Leading individuals within section E (and section C) did not support what was proposed for them. Geography's officers recognised that the case for a continuing separate geography section depended heavily upon awareness of its scientific status, itself at risk, given the large number of general papers. 'It is true,' wrote Keltie to Howarth in January 1910, 'that there has been rather too much of the travel and popular element in Section E. in past years, and I think every effort ought to be made to prevent this. As you know, Geography has a good many other aspects, and it would be quite possible to devise a meeting of the section without a single travel paper, in the old sense of the term, being introduced.'[6]

In order to see where sectional and scientific affiliations then lay, and might in future lie, a committee was established to consider the possible reorganisation of BAAS sections and to document the activities of the sections and the degree of scientific exchange between them. This committee looked at such matters for the period 1885–1909 within the BAAS and sought comparable evidence from the French, American, Italian and Australian associations for the advancement of science. To judge from BAAS cross-sectional activity (Table 7.1), geography was far from being the most active, having joint meetings with only five other sections in this period, twice with geologists and with the economists: most sections had more research committees and a larger academic programme in any one year. And elements of what geography did deliver were still based, as Keltie knew, on exploration and travellers' narratives.

These facts were used by others in the BAAS to attack the subject's intellectual rigour and the section's scientific status. Practitioners in other sections spoke against geography's claims to be a science. From Oxford, Herbertson wrote to say that he had heard that members of the BAAS committee had been sent a letter from a member of the geographical community, 'stating that there are two Schools of Geography, the Davis School and the Mackinder School, that the Davis School represents the scientific geography and is the kind of geography which is carried out in Germany'.[7] Reference is here being made to Davisian geomorphology, on the one hand, and, on the other, to the geopolitical and regional work favoured by H. J. Mackinder.[8] Keltie was much agitated, writing to numerous BAAS geographers about their perceptions as to geography's content and about the solution proposed, a move which would have restored the subject's alliance of the 1830s in seeing geography as scientific through its alliance with the earth sciences. Writing to Howarth in May 1910, Keltie was of the view that the intentions of the 'Re-organising Committee' as he termed it would not be welcome in this respect: 'I do not think that those who have done so much in recent years to advance and raise the status of Geography in this country, and make it practically an independent subject, will approve of this at all.'[9] Howarth's response the following day outlined what was being proposed:

Table 7.1 Statistics of BAAS section activities, 1885–1909

Section	Approx. No. of research committees	Total No. of papers and discussions	Average No. of papers and discussions per year	Joint meetings between sections, by section (and No. of meetings)
A Mathematics and Physics	60	1,387	55	B (4), E (1), G (1), I (1), L (1)
B Chemistry	42	749	30	A (4), E (1), G (1), I (2), K (4)
C Geology	57	916	37	E (2), H (2), K (1)
D Zoology[a]	33	407	27	H (1), I (2), K (5), L (1)
E Geography	20	573	23	A (1), B (1), C (2), F (2), L (1)
F Economics	11	534	21	E (2), L (1)
G Engineering	11	572	23	A (4), B (1), L (1)
H Anthropology	50	817	33	C (2), D (1), I (1), L (1)
I Physiology[b]	28	285	24	A (1), B (2), D (2), H (1), K (1)
K Botany[c]	18	515	34	B (4), C (1), D (5), I (1), L (3)
L Education[d]	11	155	17	A (20, D (1), E (1).F (1), G (1), H (1), K (3)

Notes:
[a] For fifteen years only, 1895–1909.
[b] For sixteen years only, 1894–1909.
[c] For fifteen years only, 1895–1909.
[d] For nine years only, 1901–1909.

Source: BAAS, Committee appointed to consider Reorganisation of the Sections, RGS Correspondence Block 1881–1910.

In reply to your letter of the 10th if I rightly understand the views of the Sub-committee, it is by no means suggested that either of the subjects, Geology or geography, should form a Sub-section of the other ... What I believe to be the suggestion is that the two subjects should form one Section which might, at discretion, for two days each meeting, divide itself into two departments.[10]

Keltie, the only geographer on the reorganising committee, was supported in his and others' views – including by the geologists – that the proposed move was not welcome.

What the debate about reorganisation revealed, in addition to perceptions of difference within the geographical community, were strongly held views from others within the BAAS about geography's status as a science. Keltie's letter to Herbertson in the wake of the May 1910 committee meeting reveals an anxiety born in part of personal shortcomings:

> I am sorry that I was alone at the British Association meeting on Thursday. . . . Armstrong [a chemist and a member of Section B] made a violent attack on Geography, as being a patch-work taken from nearly every other science, while Major McMahon [a mathematician: member, Section A] threw it in our teeth that last year, it was with the greatest possible difficulty that he could get anyone to act as President, Vice-President and secretaries of the Section, and that they appealed to me in vain, and that he did not believe that there was a sufficient number of scientific geographers in this country to supple [sic] officials for the Section.
>
> I am a very bad hand at speaking on most occasions. I did my best, but I was very much handicapped. The geologists I may say were quite with me. They abhor the idea as much as we do of uniting the two sections, but we have got to look to our position very seriously. . . . . I only wish that at the next meeting either you or Mackinder or both will be present. The discussion got very complicated and I do not know where the matter will end, probably in consideration of the entire re-organisation of the Association, its aims and objects and methods.[11]

Keltie's admission of failure in promoting geography – 'I am always a bad hand at speaking before an audience and I was really so astounded at the attack made by Armstrong at that Committee that I was at a loss what to say' – was remarked upon by W. A. Herdman: 'the only thing I can recall which could be called an attack on geography was the speech made by Armstrong to the effect, I think, that Geography was not a science, but was little bits snipped off neighbouring sciences. I expected you to get up and demolish him, but you remained silent.'[12] Herdman supported the debate: 'the officers are merely trying to solicit opinion from the representatives of all the sections in answer to a very strong and growing demand for an enquiry as to whether reform and re-organisation of allied sections would not be an advantage to the Association as a whole'. His concern with geography lay less with its scientific status than with regular attendance in section E from geography's professionals: 'My only ground of dissatisfaction with Section E is that we have had so much trouble lately in getting you distinguished Geographers to support it and *attend* the meeting *with regularity*, as the members of most other sections do.'[13] As a practical chemist, Armstrong's view of geography was based on his concern over its lack of distinctive methods: he had been arguing such of the sciences in general since the 1880s, and was a leading advocate of the view that science was not about fact but about method and process.[14]

BAAS reorganisation came up for discussion at the 1910 Sheffield meeting. Even by then, however, it was clear that not all the proposals would be accepted. From late 1910, compromises were reached: section A was renamed 'Mathematics, Physics and Astronomy (including Cosmical Physics)'; meetings' programmes were to be attentive to local circumstances wherever possible – how that was to be done was left to the discretion of local organisers; joint sessions were to be given where sensible. Although the proposed merger of geography and geology was not acted upon, the matter was not resolved: 'The Committee, whilst not prepared definitely to recommend the combination of Geography and Geology – or of any other two sections now distinct – is of opinion that this question should receive further consideration from the Council and from the General Committee.'[15]

The proposals of 1910 and their limited outcomes disclose a certain indecision in the BAAS. As Yeo has pointed out, however, what may seem indecision about sectional identity was part of wider concerns within the BAAS about the educational purpose of the sciences and their application to social life. From such concerns section L (education) was established in 1910. This section, apart from objectives in pedagogic technique, was inspired by the desire to take 'a scientific view of things in every department of life'. For Yeo, 'This was one of the contexts in which sociology, anthropology, psychology and archaeology claimed the title of social sciences and cited the use of empirical and experimental methods as evidence of their "scientific" status'.[16] So, too, was this the case for geography but in different ways. Whilst geography was striving to be seen as a social science in these terms, the reorganisation debate as it proposed to affect geography reveals concerns within the subject's ranks about the scientific focus and purpose of section E, related worries about the involvement of university geographers in the management of the section; and, if from a few, the view that geography was not a coherent science. These matters were soon brought into sharper relief.

Portsmouth 1911: geography a science? Close's 'geographical heresy'
The presidential address to section E's programme in the September 1911 meeting at Portsmouth was given by Lieutenant Colonel Charles Close (Figure 7.1). Close, after distinguished service mapping on the North West Frontier and action in the Boer War and from a previous position as head of the Geographical Section in the War Office, had taken up the post of Director General of the Ordnance Survey only days before.[17] His Portsmouth paper was in two parts. In the first,

The 'crisis' in geography   205

Figure 7.1 Photograph of Charles Close, president of section E in 1911

Close examined 'the purpose and position of geography, with special reference to its relation with other subjects'. In the second, 'what the Government, as represented by the great Departments of State, is doing for Geography'.[18]

Overall, Close's concern was to defend geography, but he did so through a particular focus upon maps and mapping, with maps understood as distributional tools for the natural sciences (he cited botany and geology in making his point) and as tools useful in the politics of Britain's imperial governance:

> It is, in fact, not only our manifest duty to encourage the systematic mapping of the world in which we live, but we should do all we can to ensure the perfection, and suitability for their special purposes, of the maps themselves. In the surveying of the earth's surface and its representation by means of maps we are treating of matters which are essentially and peculiarly our own.[19]

Most of his talk was taken up with what geography was and, more especially, with its scientific focus. His aim was the subject as a whole, not alone section E. Taking as his evidence the 296 papers published in the RGS's *Geographical Journal* from 1906 to 1910 and using the subject categories employed by the 1908 International Geographical

Congress in Geneva, Close argued that the bulk of what geography did was exploration and mapping: this was apparent in the fact that these subjects took up 57 per cent of all papers in the *Journal* over this period. Other subject specialisations each took only a small fraction; apart from 'Explorations' only 'General Physical Geography' reached double figures – 10 per cent.[20] Close further claimed that the Royal Geographical Society served best as a popularising institution and 'as a common meeting-ground for vulcanologists, seismologists, oceanographers, meteorologists, climatologists, anthropologists, and ethnographers'. Papers on these topics could easily have found a home in specialist organisations and scientific journals rather than in geography, Close argued. 'If we thus study the relations between geography and other subjects,' he continued, 'we are almost bound to arrive at the conclusion that geography is not a unit of science in the sense in which geology, astronomy, or chemistry are units.' Although he recognised that other subjects drew upon geographical work and that there was, as we have charted in detail, an emerging trend of thematic human and regional geography, Close ended by questioning geography's scientific credentials altogether: 'Before this human aspect of geography – or, for that matter, any other aspect of the subject – is recognized by the world of science as an independent, indispensable, and definite branch of knowledge, it must prove its independence and value by original, definite, and, if possible, quantitative research.'[21]

Close's claims may have been founded on limited evidence, and we may conjecture that his promotion of mapping was motivated by personal interests, even, perhaps, by his recent appointment to a position of national importance. He recognised that section E provided a 'general meeting-ground for all workers in the various divisions of earth-knowledge' and that other sections 'incorporated geographical work' but still felt that 'geography, in the sense mentioned, is not so much a subject as a method of research'.[22] But if his tone was intemperate, his insistence that geography was not a distinctive science was unsettling to his peers. As a whole, the evidence served in his eyes less to demarcate what geography was and more what geography ought to be and why it was failing to be scientific and was seen to be so by non-geographers.

Public reaction was muted. The *Times* reported:

> It has been evident for some years that the section devoted to geography has been undergoing a change. The crowds who used to throng the section room to see and hear distinguished travellers are no longer to be met, the number of travel papers has greatly diminished, and old frequenters of the section have been heard to complain that the interest in geography is dying out. As a matter of fact the reverse is the case. The place of the traveller has

been taken by the worker at geographical problems; the diminished audiences consist far more than they did of geographical teachers and students, and if the papers presented appeal less than they used to do to the outsider the discussions have become fuller and more authoritative.[23]

Professional commentary upon the Portsmouth meeting omitted mention of the nature of Close's address.[24] Private reaction to what Clements Markham described as Close's 'geographical heresy' was, by contrast, considerable and prolonged.[25] Keltie wrote to Close to express his reservations. Close was unrepentant and forceful in response:

> If you and those who think with you will try and make Geography a little less indefinite and if you can persuade the world of science that Geography is doing scientific work of a kind which is not being undertaken by other agencies, and if you can frame a definition of the subject as a science which will be accepted by representative scientific bodies such as the R.S.[Royal Society] or the B.A. [BAAS], I for one shall be glad. But it is no good crying peace when there is no peace. The position which Geography at present occupies in the scientific world is really humiliating. . . . It is outside scientific men we must convince.[26]

For Markham, Close's address was 'perfectly astounding', and 'must be challenged'. George Goudie Chisholm, lecturer in geography at Edinburgh, rebutted Close by pointing in his teaching to Ellen Churchill Semple's *Influences of Geographic Environment* (1911), whose brand of cultural and environmental determinism Chisholm saw as signalling a new focus for geography, and to recent work by the French regionalists. J. W. Gregory, a geologist at the University of Glasgow who taught geography there, offered a placatory interpretation, noting, 'I thought all Close meant was that cartography is the primary duty of geography', and that he (Gregory) 'did not think it [Close's address] would have the slightest effect on the British Association or on the development of geography in this country'. Keltie thought Close had been put up to it by McMahon, one of the secretaries of the BAAS, 'who for some reason is inimical to Geography' (and who, as we have seen, spoke against the subject in May 1910). In fact, it is likely that McMahon spoke as he did because of Close's backing – for it was Close who had written during the reorganisational debate in 1910 to the effect that geography was split between two constituencies.

> There are two schools in the geographical world. One which may be represented by Mackinder who says that geography is not a science but an 'attitude', and one which may be represented by the American Professor Davis who says that geography is 'Modern geology'. Most of the Mackinder men are unscientific, and use the term geography as a convenient expression to

cover certain minor studies which they call political geography, military g., economic g., &c. To my mind one might just as well talk of political chemistry; they arn't [sic] bad little studies but are in effect the application of certain geographical facts to politics, war, economics, &c. Now Davis makes the main business of geography the study of the changes of the earth's surface which we can actually watch, and the explanation in geological terms. This is I believe a living and useful branch of science. None of the Mackinder school have produced any generalizations. If I were asked to expand Davis's epigrammatic definition I should say that geography might conveniently be divided into physiography (in Huxley's sense) topography (exploration &c.), and geodesy.

For these reasons, concluded Close:

> I am therefore very much in favour of anything which tends to ally geography with geology and astronomy (the latter on the geodetic side), but especially with geology.[27]

By late September 1911, Hugh Robert Mill, George Chisholm and A. J. Herbertson had met to discuss the matter. Mill confided to them that, in being present at the Portsmouth meeting, it was he who had effectively limited the public response to Close's address by inducing 'the *Times* man to suppress the controversial part'.[28] Collectively, what concerned these and other correspondents was that Close had denied geography scientific status and abused his position as president of section E to do so. Close's connection with McMahon was not commented upon (although, in a letter to Keltie of 11 May 1910 on the correspondence over two schools of geography, Herbertson had written, 'I can guess at the author of this communication.'[29]) At just the moment geography's intellectual position within the BAAS as well as the RGS was shifting further from exploration to 'scientific geography', its sectional identity still under review and geography's credentials yet to be determined precisely even as the subject was becoming professionalised, Close's words seemed to threaten the very status of the subject.

Yet nothing happened, at least publicly. Despite their deep-felt dismay, those involved in the BAAS and in the RGS took no public action. Following guidance from Lord Curzon (RGS president), amongst others, Keltie wrote to all concerned, 'The President has decided that the best plan is to take no official notice of Close's Address – simply to publish it in the [*Geographical*] *Journal*. If it leads to correspondence, good and well, if not, it does not matter. Close is now beginning to hedge a bit!' He wrote to Mill to report on the general sentiment:

> Both the President and Darwin and Johnston are of opinion that the best thing to do is to take no notice officially whatever of Close's Address, but

just to print it in the Journal. This they think would be much more effective than if we put ourselves in the dock to defend ourselves against a baseless accusation. If it leads to a correspondence good and well, let it do so. I do not quite understand what you mean by the Society rewarding amateurs and disregarding the serious geographer. I wish you would speak more plainly.'[30]

Markham dissented – 'Lord Curzon may be right as a question of policy, but I think not. We *are* in the dock.' William Morris Davis echoed Curzon: 'Your view of Close's paper is right. It is better not to magnify its small importance.'[31]

Of small import or not, Close's 'heresy' identified fault lines across and within geography and not just within the BAAS, fault lines that had to do with geography's scientific focus and perceived lack of epistemological unity or distinctiveness in method other than in mapping. The decision made by members of geography's emergent professional community (many, leading figures within the RGS and BAAS) not to go public against Close should not obscure the fact that, privately and collectively, the Close affair was a matter of great anxiety. If heresy it was, then this public pronouncement and the private disquiet about it hint at presumed normative views regarding what geography was or, perhaps, should have been as a science. Yet the evidence for what sort of geography was being undertaken in and through the BAAS and how geography in the BAAS was seen by others suggests no such normativism existed in practice. For Close, this was weakness – 'modern' geography was still too strongly exploratory, without a compelling case to be termed scientific other than as a form of geology. For others, geography was an 'attitude', a method more than a science.[32]

Dundee 1912: absences, audiences and local geographies
What these events also disclose, certainly in private correspondence, is that not all geography's professionals saw section E as a forum either demanding or deserving their attention and that BAAS sectional officers did not always work together. Herdman's remarks in May 1910 about the lack of attendance from professional geographers at section E meetings were echoed in exchanges between Keltie and Howarth in September 1912 as they reviewed the just-passed Dundee meeting. 'There is no doubt,' Keltie protested, 'that an effort ought to be made to get the younger geographers to throw themselves into the work of the section.'[33] Howarth noted that the section 'doesn't attract them at present'. William Barton, recorder to the section, put it more strongly: 'the dissatisfaction with Section E is fierce – from within this time! The list you give of non-attendants [referring to the 1912 Dundee meeting] is itself evidence enough of rottenness. We *must* have a president who will

command their confidence and rope them in.'[34] Keltie's reply noted that he thought the problem lay outside section E – 'I was under the impression that the dissatisfaction with section E. existed among the Officials of the Association' – yet admitted that more needed to be done within the section: 'But certainly we ought to enlist all the young geographers to support the section in future.'[35]

These within-section differences were heightened, as Keltie and Howarth knew, by the circumstances surrounding the planning and the running of the 1912 Dundee meeting. Responding to BAAS general directives that sectional programmes reflect the circumstances of the host town, several of those persons planning the Dundee meeting determined to use the recent strategic directives to advance a scientific programme incorporating local interest and topics of wider importance, but to snub English-based RGS-influenced section E committee members and to energise the moribund Royal Scottish Geographical Society in doing so. For Robert Rudmose Brown, lecturer in geography at the University of Manchester, writing in April 1912 to his erstwhile Antarctic colleague the oceanographer William Spiers Bruce – the two had sailed together on the *Scotia* as part of the 1902–04 Scottish National Antarctic Expedition[36] – Dundee should be a Scottish affair:

> It is not a matter of national sentiment – at least to say so would be unwise: it is a matter of national capacity & intellectual standing which must be voiced by a scientific society such as the R.S.G.S. Of course a local (Dundee) section secretary will have been appointed, but I don't know who he is. He should get as far as possible papers dealing with Scotland & Scottish work & do his utmost to force them on the committee. . . . We must meet the English injustice with work & not with sentiment.[37]

Attempts to enlist the Earl of Stair, president of the RSGS, and others to the Scottish cause were unsuccessful. And so, Bruce wrote to Brown:

> It rests practically with you and myself to make the Geographical Section of the British Association as thoroughly Scottish as we can at the meeting in Scotland. The Council of the Geographical [the RSGS] are an obsolete body of old men who have no other idea than to make a lecture agency of the Society, and Bartholomew [J. G. Bartholomew, head of the map-making firm of that name and RSGS co-founder and honorary secretary] is not interested beyond filling his purse.[38]

In the event, plans to make Dundee 1912 especially Scottish and so limit the influence of BAAS London officers and RGS men succeeded only in making for a crowded programme that satisfied no one. Some papers had to be cut short, others delivered too hurriedly 'to an audience obviously balancing the claims of lunch against additions to their geographical knowledge. As time went on the scale went down more and more against the latter, and the exits were frequent and disturbing.'[39] Markham – 'that old fool and humbug' as Brown privately termed him – gave offence by emphasising Captain Scott's Antarctic endeavours and 'giving scant justice to other explorers', i.e. Bruce, Rudmose Brown, the Scottish National Antarctic Expedition, and even omitting to note the fact that Scott's ship, the *Discovery*, was Dundee-built; Watson's presidential address ran over time; Close's paper on world mapping prompted an ill managed lengthy discussion: in sum, Dundee 1912 was, for section E, a disaster.

Poor management for particular reasons could not disguise general doubts over scientific purpose. As an anonymous commentator noted:

> There were certain defects in the programme which might well engage the attention of the officers. It seems clear that it is time that some more definite decision is come to as to what constitutes geography. Broadly speaking, it may be said that the British Association has two objects in its sectional meetings – to permit specialists to exchange views and give the results of their work, and to offer the general public an opportunity of learning the kind of contributions which specialists in each branch are making to the common fund of knowledge, and the line along which the individual subjects are developing at the present time.
>
> If we turn to geography we find that the sectional Proceedings in recent years suggest . . . , the prevalence of the belief that almost any subject may be regarded as geographical, though there is a distinct tendency to think that generalised descriptions of unknown or little known regions are specially geographical. . . . In our opinion, if Section E is to justify its continued existence, geographers must make up their minds to attend its meetings in sufficient force to impose their own point of view, and to stamp out the idea that the Section is the happy hunting-ground of the collector of interesting but unrelated facts, in a word – of the untrained and un-specialised observer.[40]

This difference between specialists and generalists in BAAS science as a whole and in geography in particular ran deep. But what the events of 1910–12 show is that the fault lines striking through and across geography and the BAAS did so in several directions. Was geography a science or not? For some scientists outside section E – and, crucially, for some within it – the answer was no, or, at least, not science enough. Geography had neither method nor an explanatory focus it could call its own. It offered only distributional emphases in mapping and synthesis of other sciences' work. Was 'the geography community' one thing or not? Practitioners seem to be divided within section E and, more generally, between professionals and 'distinguished geographers' who, no matter that they might be advocates of Davisian geomorphology, Mackinderesque geopolitics, Closian cartography or other emergent thematic specialism, were sufficiently disenchanted with the section by the early 1900s that they did not attend its meetings in sufficient numbers. Populists drew the crowds but hindered the science.

Some things did change for section E in the wake of Dundee 1912: the 1913 Birmingham meeting was the first to break the dominance of its secretariat by RGS officials; younger professionals were prevailed upon to give research-led papers; of the seventeen papers given there, only two were 'exploration' in type.[41] Nevertheless, in turning to examine in more detail how geography developed as a science of education and of empire in the twenty years after 1919, it is important to bear in mind that neither as a subject, science or not, nor as a community, professional or not, was geography a shared and agreed-upon thing. This is not to suppose that it should have been, but it is to point to the several ways in which geography and geographers were at work, being judged and combining with others in and outside the BAAS.

World citizenship and local science: debates and practices in geography, 1919–1939

The 'thickening' of the BAAS archives for the two decades after 1919 attests to how the Association was then increasingly engaged, alone and in combination with other bodies, in promoting science to the public and in co-ordinating its activities to ensure science was a social good. Papers survive on BAAS science and the civil service, on the workings of a BAAS committee constituted to consider afresh 'the advancement of science to social progress' – defeated by the scale of its task, the committee soon disbanded – on correspondence over the Association of Scientific Workers, connections with the British Science Guild (which body merged with the BAAS in 1943), liaison with organisations such as the Council for the Preservation of Rural England and the Royal Society, to name just two, and, for various periods, the minutes of the Conjoint Board of Scientific Societie, of which the BAAS and the RGS were members.[42]

In these terms, MacLeod's description of the post-World War I period as marked both by an intensification of debate and by a degree of disenchantment within the public over what science and the Association was about, how it should work and for whom, strikes more clearly true as debate than disenchantment.[43] Against the backdrop of such debate and heightened activity, two illustrations of geography's concerns and purpose in the period connect not only with the longer-run place of the subject but also with some of the proposals of 1910 – namely, that sections should work more closely with one another and that meetings should better reflect their local setting.

Geography, education and world citizenship

We have seen how the BAAS overseas meetings were from 1884 significant in promoting the connections between science and empire and how, after 1887, geography was developed as a subject in British education, in universities and in schools. Although anxieties had been expressed before 1914 over the educational bases of imperialism, education for empire became an important feature of the BAAS mission after 1918 because of a World War which had exposed Britain's dependence upon and responsibilities towards its dominions. Reparations included a return to the belief that science transcended national boundaries and could unite the League of Nations anew. Geography as a subject lent itself to international collaboration and, in Britain, to imperial renewal. At the 1919 Bournemouth meeting, for example, T. W. F. Parkinson, secretary of the Manchester Geographical Society and an advocate for the further strengthening of geography's place in the educational system, spoke to the connections between geography and empire:

> Probably there never was a time in the history of man when geography was so necessary as at present. . . . We have the largest empire the world has ever seen. . . . Unless we know more of this empire . . . we are unworthy of the trust which is imposed upon us.
>
> How can we expect the respect of Canadians, Australians, South Africans, and Hindus when we know so little of them and their countries and make such small efforts to know more? . . . Surely it is necessary to study geography if a right conception of our obligations is to be obtained.[44]

Yet whilst Parkinson's words struck a sympathetic chord with many in the BAAS hierarchy, several geographers present demurred from his imperial emphasis, since, as C. B. Fawcett put it, the fundamental unit of geographical study was the world not the British Empire. What everyone did agree upon was the still poor role afforded geography teaching, in British schools and for imperial citizens alike.

One expression of this concern was the establishment in 1920 of a National Committee for Geography, managed jointly with the Royal Society. The remit of this BAAS-dominated committee, which operated until at least 1952, was to provide an overview of geography's role as an educational subject in the several sectors in which it was by then operating.[45] Following a resolution advanced by H. J. Fleure, section E from the early 1920s allied with the educationalists of section L to advocate a greater place for geography in advanced courses in school.[46]

A further and related expression was the establishment of a joint committee under the chairmanship of the educationalist T. Percy Nunn. This committee made repeated claims for financial support between 1922 and 1928 (see Table 3.2). Nunn's committee included the geographers H. J. Mackinder, W. J. Barker, H. J. Fleure, O. J. R. Howarth and J. F. Unstead amongst its members. Writing to the Board of Education in March 1921, the BAAS general secretaries laid some of the groundwork for the Nunn committee by repeatedly stressing how, in its present place in advanced studies, geography was neither in the 'science group' nor in 'modern studies'. Yet, because 'it has been one of the effects of the War to widen the public outlook on geographical matters, and to increase the demand for geographical problems', it was argued that 'no avoidable hindrance should be allowed to exist, to the provision of advanced courses in geography, and to the studies of potential teachers and students of it'.[47] It was up to the Nunn Committee as a whole as to where they wanted to place geography – science or not – and correspondence on this question is revealing. 'In a certain sense,' Nunn wrote to Barker, head of geography at the University of Southampton, 'the discussions of our Committee have moved around the idea that Geography is an autonomous science, not a collection of snippets from other sciences'. Close's denunciation was for some still fresh in the mind. As Nunn further observed:

> On the other hand, if the Board is asked to recognize Geography as a Science, there is reason to fear that they may regard it in the second light not the first. I think, therefore, that Professor Myres should make clear that in asking the Board to admit Geography among the science[s] he is not asking them to admit a modern edition of the old physiography, but a discipline which claims the name of Science on the ground of its definite and characteristic form of unity.[48]

A final decision as to whether, for the purposes of British secondary school education, geography was a science or a topic of modern study was never arrived at. The Board of Education refused to judge one way or the other. To do so, it argued, would remove the responsibility from individual exam boards and school teaching programmes, and the Board was not prepared to do that. Whilst it is the case that the Nunn Committee took an explicit position on geography as a science, it is also true, as Collins has shown, that it implicitly accepted the view of Cyril Fawcett and others that geography should take a world view and not an empire view. More than that: the fundamental role of geographical education was international understanding and the promotion of international (rather than imperial) citizenship. To that end, officers of the

BAAS worked with those of the Geographical Association more than with those of the RGS and did so because, as individuals, they shared the same view of a subject whose teaching would lead to better international understanding: in these terms, geography was a form of global civics. As Percy Roxby put it in his 1930 presidential address, 'We may claim for human geography that, rightly studied, it is a vital element in training for national and international citizenship.'[49] His internationalist agenda was supported in later years by Fleure, for example, in 1932 and Lord Meston in Leicester in 1933.

Others within section E disagreed, some vociferously. Geography's educative role should be, they argued, explicitly imperial, evidently British. Such was the view of Vaughan Cornish, and he put it stridently in his 1923 address in Liverpool (see also Winterbotham 1936: table 3.4). Entitled 'The Geographical Position of the British Empire', Cornish disagreed with the internationalist agenda being proposed for geography by Nunn, Barker, Unstead and others, but not with their view that geography could contribute as a science. Many in section E that year spoke to the local import of their subject – and were commended by the Liverpool newspapers for making 'a special feature of local interests'.[50] Cornish looked to bigger concerns altogether. His talk was part survey of the British Empire's strategic position in respect of world transport and communications, part polemic about the undesirability of allowing Britain's colonies to be flooded with immigrant labour – 'Enough is known of the relation between geographical environment and national well-being to declare with confidence that the decision to erect a Barrier against coloured labour in tropical Australia is best both for the white race in Australia and for the coloured people of the monsoon region of Asia' – and part eugenicist manifesto.[51] Cornish advocated 'the ideal of a family of four persons' (amongst the professional classes only, even whilst noting that the working classes made hardier colonists) and the management of birth and death rates in order to ensure that 'the British population in the Dominions exceeds that in the Mother Country'. 'The maintenance of the race will then rest chiefly with our people oversea [sic], and, with their great resources, it should be possible for them to keep pace with the other growing nations.'[52] As he put it, 'Therefore in our own interests and in that of the coloured races (who conflict amongst themselves) it is desirable to maintain the present proportion of the British stock, to whom the Empire owes the just administration of law and a progressive physical science.'[53]

Whatever his peers thought to it, none thought to restrict its coverage by the press as they had for Close. Cornish's paper prompted considerable public response. *The Evening Standard* supported his views. *The*

Scotsman deemed his talk 'Malthusian'. *The Daily Herald* reported on Cornish's imperialism as 'a kind of lunacy':

> Many members of the British Association who listened yesterday to the address of a certain Dr. Vaughan Cornish, in the Geographical Section, must have been surprised at its blatant Imperialistic tone. In the strident voice of a drill sergeant on parade he orders all British parents to have at least four children. . . . the British People in this island are not Robots. They mean to control their own lives, they mean to make Britain a land fit for them to live in. They will emigrate if they feel like it, but they will not be treated as pawns for crazy Imperialists to shift about as they please.[54]

The Daily Telegraph was more measured:

> We look to the British Association for advice and instruction on many dark problems and are seldom disappointed. Yet it is with a certain surprise that we hear the president of the Geographical Section telling us how many children we ought to have. It is a matter upon which we should hardly have inclined to allow geography the last, decisive word.[55]

Nor is it my intention here. As these examples illustrate, geography was made to serve different intellectual agenda for the BAAS in the inter-war period – here, global internationalism and British imperialism – even as other practitioners promoted more specialist themes, had doubts about the subject's claims to be a science or saw divisions between physical and human geographical study and no essential subject unity. At the same time, however, the emergence of regional geography did offer a means by which the BAAS could turn to geography and to geographers in order to serve its strategic mission.

Writing science locally: Association handbooks and regional survey
Writing in June 1913 of that year's BAAS Birmingham meeting, a reporter for the *Birmingham Post* clarified for the reading public the nature and purpose of the meeting's publications:

> The custom of the Association is to induce the local committee of the place of meeting to publish two handbooks, one for the enlightenment of the visiting member, and enlarging upon the history, topography, organisation, and scientific interests of the locality; the other for the enlightenment of the local member who, in nine cases out of ten, knows little of his or her neighbourhood. The first is the handbook; the second is the excursion guide-book. The handbook is a work of reference, a volume of some 500 pages, laborious and expensive to produce. The guide-book is a small pocket affair that can be easily carried and consulted. Both of these books are given free to every member or associate on presentation of their tickets at the reception room.[56]

This distinction was not always evident in practice. For all that it was a text promoting science's local dimensions (see Chapter 2 above), the BAAS handbook took over forty years to emerge as a distinctive expression of scientific activity suited to local circumstances. The first handbook to consider the local geography and history of the location in question was produced in 1874 for the Belfast meeting, under the direction of the Belfast Naturalists' Field Club.[57] Its content embraced the region's physical geography, geology, botany, zoology, topography, history, antiquities, agriculture and trade and commerce, with a short section on the excursions. The 1874 Belfast volume effectively established the model for later handbooks, with the exception of the overseas meetings, whose handbooks, often longer and in more than one volume, had, as we have seen, a strongly imperial and utilitarian agenda, portraying the colonies to the visitor and stressing science's importance to the development of Britain's overseas dominions.

In their chronological focus, thematic examination and synthesis of regional character of an area, BAAS handbooks foreshadowed elements of the regional method as espoused by some early twentieth-century geographers. Knowledge of place in this form meant that science was invested in the local hosts rather than in visiting 'experts': textually as well as via excursions, BAAS science was made in its local provincial settings, not imported to them. Regional knowledge from the bedrock to the latest facets and sequences of human occupance in the historical present helped provide a method. Local science in this form could be accumulated, worked with to aid local political as well as imperial or internationalist agenda. Later BAAS visits to the same town afforded a chance to revise errors in previous handbooks – organisers in Glasgow in 1901 looked upon the three volumes produced as 'an opportunity for repairing many of the errors and omissions of the former handbook and for bringing to a focus the very large amount of work which has been done of late years in the Clyde area'[58] – or to publish what existed only in manuscript or as field notes held by local specialists, as was stressed of the Southport volume of 1903. In keeping with the intentions of some organisers, the 1912 Dundee handbook aimed not just at the temporary edification of Association visitors but at the longer-term benefit of local citizens, that they might be 'further strengthened in their feelings of local patriotism and towards endeavour to raise the standard of life and thought in our community'.[59]

From the 1920s, BAAS handbooks began to establish definitive standards for the presentation of science in provincial context. Where the 1839 *Queries* and the 1874 *Notes and Queries* had sought to instil methodological procedures for the investigation of human difference, handbooks in the 1920s aimed at regional description based on geographical principles. They did so as a result of links between the Conference of Corresponding Societies and developments in the nature of regional survey within geography, mainly from within university geography departments.[60] The view of delegates to the Conference of Corresponding Societies that met during the 1925 Southampton meeting was that more needed to be done 'to secure the establishment and [to] facilitate the extension of regional researches, especially in the districts which it [the BAAS] visits'.[61]

It is unclear whether this proposal was in direct consequence of the directives in 1910 to effect closer links between BAAS science and meetings' venues, but it was certainly consistent with them, as, of course, it was with the longer-run practice of connecting sectional programmes to local agenda. This proposal was adopted by the BAAS General Committee and realised through collaboration between section E and the Geographical Association, whose own Regional Survey Committee was then helping advance the study of regional survey in Britain.[62] As the agricultural scientist Sir John Russell put it to historian and archaeologist Sir John Linton Myres, such regional survey work was best undertaken in association with departments of geography in local universities: 'the best way of putting the principle into practise would be in each case to ask the University School of Geography to undertake the work. It is the kind of thing that many of them are doing and all of them ought to do.'[63] The 1922 Hull handbook was taken as the model for BAAS handbooks as a distinctive genre of regional text (by its compiler at least):

> My modesty will no doubt prevent me [noted T. Sheppard, director of Hull's municipal museums and editor of the 1922 Hull handbook] from stating that the Hull Handbook might be taken as a model for future places. All I can say is that Nature and other Journals stated that it was as near perfection as it could possibly be for a Handbook of this sort. Possibly had it had 'Regional Survey' IN BIG! letters on somewhere it might have been absolutely perfect.[64]

Regional handbooks did not necessarily require new science. As Myres put it, 'I regard the current suggestion about regional survey rather as a change in the mode of preparing for a B.A. meeting, than as a demand for a new handbook on every occasion. Where there is a good Handbook in print, it should be adopted & supplemented (as I understand Oxford will do): where there is none, the B.A. offers I think a real service in proposing to co-operate in producing one.'[65] For him, 'The Handbooks have always been a *pictas* offered by the Local Committee to the visitors. They have been all sorts of shapes, thicknesses, designs and executions. Sometimes an existing local guide has been used, or re-edited for the occasion, and supplemented by a volume of essays on some local interests.'[66]

Rather than that for Hull in 1922, the handbook that signalled best these close connections between section E, university geographers and the localist agenda for BAAS scientific meetings was that produced for the 1927 meeting in Leeds (Figure 7.2). This was produced under the editorship of Charles Fawcett, Reader in Geography at Leeds and a leading advocate of regional studies and regional method in geography. In his 1919 *Provinces of England*, for example, Fawcett had identified the importance of regional studies to civic identity and as a basis to future planning, and in the BAAS 1927 *General Handbook* he contributed chapters on the city's location and its 'home area'.[67] He had been planning such work since at least 1925: letters to Myres in December that year refer to 'a considerable amount of work done towards a regional survey', including several map studies, MA theses by Leeds geography students – two of which were to be the basis of papers at the Leeds 1927 section E lectures – and population surveys 'which agreed with the limits I drew in my "Provinces of England"'.[68] This was a form of local geography writing for broader educational ends – local geography and provincial science as a form of civics and citizenship – and was intended as a model for later work through which local geographies would present the whole country:

> The effort was further made at Leeds to organize the preparation of the B.A. Handbook so as to stimulate, in advance of the meeting, all local enquiries of this kind, and arouse a keener interest in local features and activities, focussing such studies on the projected Handbook, but hoping to perpetuate this kind of work in each district, as it was visited by the B.A. in its turn, until the whole country was covered by a series of such volumes, which it was hoped might approximate to a standard format, of which the main features were suggested by the best of the previous Handbooks.[69]

After 1932 the BAAS handbook was produced in standard format as a *Scientific Survey* of the region in question – testimony to the importance of regional geography and the work of Fawcett, Myres, Roxby and others in advancing geographical methods to the good of the BAAS as a whole.[70] As Stevens noting in opening his address to section E in 1939, 'The most conspicuous geographical concept whose enunciation is associated with British geographical thought is that connected with the phrase, *Natural Geographical Region*.'[71]

There is another sense, however, in which we should read geographers'

British Association
Leeds Meeting 1927

GENERAL HANDBOOK

Edited by
C. B. FAWCETT, B.Litt., D.Sc.
Reader in Geography, University of Leeds

Published for the Local Committee by
CHORLEY & PICKERSGILL LTD
THE ELECTRIC PRESS LEEDS
August 1927

W16-8484

Figure 7.2 Productions in regional science. The cover of the BAAS Handbook, Leeds, 1927. With this work Fawcett and colleagues in the Leeds geography department established a model for regional science and BAAS handbooks

involvement with the BAAS handbooks in the late 1920s and early 1930s as more than the use of regional methodology to revive and, later, standardise one form of the Association's public mission. The BAAS had no publication series, although the annual report carried abstracts and the presidential and sectional presidents' addresses. Papers given in its meetings usually had to seek publication in specialist journals. For most British geographers this meant the *Geographical Journal*, the *Geographical Teacher* (the first associated with the RGS, the second with the Geographical Association) or the *Scottish Geographical Magazine*, this last under the editorship between 1902 and 1934 of Marion Newbigin, a leading figure within section E and advocate of the 'new' human geography including regional geography.[72] But these outlets were insufficient for geography's developing range of activity and its expanding professional community. Association handbooks thus offered geographers a chance to show the BAAS that regional method could provide a form of systematised scientific enquiry to address local audiences, a development consistent with the organisation's mission.

The relative absence of publication outlets for BAAS geographers, professionals or otherwise, was compounded by what members of section E and others saw as the reluctance of the RGS to publish research papers, especially in human geography, or to support monographs in the subject. There was a sense, too, that the RGS tended to favour general description or exploration papers. Late in 1929, at the instigation of the RGS president – none other than Sir Charles Close – Arthur Hinks, secretary to the RGS, wrote to several leading figures (his choice of correspondents being guided following consultation with Fawcett) over Close's concern that 'this Society might profitably do more to advance the subject of human geography both in choice of subjects for Afternoon Meetings and by publication in the "Geographical Journal"'.[73] Hinks undertook a survey of publication trends in geographical publications in order to inform the group's discussions (Table 7.2). Papers on exploration predominated in the RGS's *Geographical Journal* – in which outlet publication in human geography had declined between 1912 and 1927–29, in contrast to the other surveyed journals – and physical geography had a strong presence over this period. The increase in human geography research publication within the *Scottish Geographical Magazine* is attributable to Newbigin's editorial efforts; the strength of historical geography within French geographical periodicals reflected the prominence of the French regionalists and Vidalian emphases upon the character of cultural landscapes examined over time. It is not clear, however, how far this evidence informed later discussions within the RGS. The RGS sub-committee formed under Close's direction met nine

Table 7.2 Categories and trends in geographical article publishing, 1912, 1927–1929

Year	Category and No. of articles published, by journal						
	Exploration	Maps and surveys	Physical geography	Historical geography	Human geography economic, etc.	Theory of education	Other
Geographical Journal							
1912	43	7	18	4	26	–	–
1927–29	46	16	18	7	9	–	–
Geography							
1912	–	5	–	–	35	60	–
1927–29	–	5	17	3	39	32	4
Scottish Geographical Magazine							
1912	23	3	23	12	36	–	3
1927–29	15	5	14	7	53	–	6
Geographical Review							
1912	–	–	–	–	–	–	–
1927–29	16	6	25	4	47	–	3
Annales de géographie							
1912	4	9	48	–	35	–	4
1927–29	–	–	17	–	74	–	3
Géographie							
1912	25	5	17	17	30	–	6
1927–29	25	3	7	13	45	–	7
Zeitschrift-G. Erdkunde							
1912	29	–	47	8	16	–	–
1927–29	10	–	57	5	23	–	–
Petermann's Mitteiln.							
1912[a]	12	3	45	3	18	3	15
1927–29	8	7	34	12	18	9	12

Note:
[a] First six months of the year only.

Source: RGS-IBG Archives, Correspondence Block CB 1921–1930.

times between December 1929 and February 1934 without coming to any firm recommendations. Noting at its final meeting that its work had met 'with a very meagre response', the then president, Sir Percy Cox, remarked only that further work needed to be given to 'removing any latent feeling which might still exist, that the Society was inclined to neglect the human and educational aspects of geography'.[74] Such feelings, latent or otherwise, did exist, or had earlier done so. Replying to Hinks's letter, the Cambridge physical geographer Frank Debenham made clear what he thought the problem was:

> Without being an opponent of human geography in any way, I cannot regard most of the descriptive stuff they favour with any enthusiasm, but I suppose that it can be raised to a place comparable with the physical side of geographical inquiries. At all events, unless it is to give us something more than enlightened guide-books I do not think the Society can give it very strong support.[75]

By February 1934, Cox's entreaties were too late. The Institute of British Geographers had begun less than a year before, citing the unwillingness of other bodies to publish research in human geography as one of the principal motives for its foundation.

Conclusion

Looking back in 1945 on a lifetime's involvement with geography and with section E, Hugh Mill looked to his own presidential address of 1933 – where he had defined geography as 'the description of the earth's surface with special regard to the forms of vertical relief and to the influence which these forms exercise on all mobile distributions' – to the foundation of teaching departments and the effect of the World War of 1914–18, which 'made everyone eager to follow events on maps' as helping establish that 'fertile soil, in which Geography is flourishing as it has never flourished before'.[76]

The evidence presented here complicates Mill's recollections of geography's more recent past with regard to its place in the BAAS. Geography was involved in the inconclusive plans of the BAAS to reorganise itself, in 1910 and again after 1919, to meet public expectations and to better manage its resources and bring subjects together. The BAAS continued to be a social affair into the 1930s even as many held doubts over its scientific work (Figure 7.3). Although geography had a history of what we might term sectional 'identity shift' before then, and of affiliation for research purposes with other sciences, it is noteworthy that the main cross-sectional collaboration between geography and other BAAS sciences after 1919 was with section L, education, and then as part of an internationalist agenda and as part of debates over whether, for educational purposes, geography should be identified as a science.

Geography's development as a subject for teaching within schools and universities in early twentieth-century Britain was not paralleled by agreement over its focus, content or methods as a science. Differences

Figure 7.3 Afternoon tea at the BAAS meeting, Leicester, 1933

were expressed over what Keltie termed its 'travel and popular' nature, and a more evidently research-led profile. Differences were identified over the human and physical dimensions of geography, over method, over emphases upon explanation or description. Differences existed between geography's internationalist or imperial agenda, if less so over its educational role. At certain moments, and from within section E as well as from outside it, the view was held that geography was not a science. The fact that Close's views startled some of his contemporaries has, I suggest, as much to do with the social position and place from which he said it – as president of section E and in a formal address – as it does with what he said. To judge from his peers, he was not alone in his views. If all he was doing was emphasising cartography as geography's 'primary duty', he was echoing the views of other notable predecessors. But to argue that geography was not a 'unit' in the same way as other sciences – with clear objects of study and agreed methods and explanatory principles – certainly cast doubt upon the subject's scientific status.

In all of this, of course, there are echoes of previous debates: about popularity, about the procedures for establishing a subject or its methods as 'scientific', about the intended audiences for geography

The 'crisis' in geography 225

Figure 7.4 Excursion science in the inter-war years. These delegates to the BAAS meeting in Norwich in 1935 are on an excursion to Cromer

within a body whose remit at foundation had been to advance science to the public but whose public had, in turn, flocked to geography sessions at the BAAS annual meetings precisely because it was seen as less a science than some other subjects. Several of these concerns continued into the modern period: writing to Mill about the forthcoming BAAS meeting in Norwich in 1935, Debenham remarked 'I would like to have a paper or two on plain exploration as well from people, but Section E has gone a little chilly on exploration of late and I do not want them to think I am altering their steady policy too much'. Even so, he continued, 'I have noticed, however, that the meetings which are best attended are those which have an exploratory tinge to them.'[77] And, as Figure 7.4 suggests, fieldwork continued to attract delegates, sustaining the link between sociability and scientific camaraderie in the field (Figure 7.4). Four years before, as the BAAS celebrated 100 years of activities, the opinion was being expressed that geography had its place – but not as a science. 'The address in Section E (Geography) [by the Oxford-based regionalist and political geographer H. J. Mackinder] would be delightful as a magazine article and ought not to be buried in the Association volume. . . . It is one of the subjects that has been cleverly talked into a hearing of late years but it is not and can never be a science – because of its universality.'[78] The author of the letter was one Professor H. E. Armstrong. Nearly a quarter of a century after his verbal declaration had silenced Keltie – a silence which in turn had caused consternation amongst geography's proponents – one view at least about geography's place as a science remained unaltered.

Notes

1 R. MacLeod, 'Retrospect. The British Association and its historians', in R. MacLeod and P. Collins (eds), *The Parliament of Science: The British Association for the Advancement of Science, 1831–1981* (Northwood: Science Reviews, 1981), pp. 1–16, quote at p. 2; R. Overy, *The Morbid Age: Britain between the Wars* (London: Allen Lane, 2009).

2 Other than that part of the work of Pancaldi and Worboys discussed in Chapter 4 in relation to the internationalist agenda of the BAAS after 1900, the study of the BAAS in the twentieth century is effectively represented only by that handful of essays in MacLeod and Collins: D. Layton, 'The schooling of science in England, 1854–1939', in MacLeod and Collins (eds), *Parliament of Science*, pp. 188–210; P. Collins, 'The British Association as public apologist for science, 1919–1946', in MacLeod and Collins (eds), *Parliament of Science*, pp. 211–36; D. Morley, 'Yesterday, today and tomorrow, 1945–1981', in Macleod and Collins (eds), *Parliament of Science*, pp. 237–60.

3 R. W. Steel, *The Institute of British Geographers: The First Fifty Years* (London: Institute of British Geographers, 1984), pp. 13, 14, 17.

4 MS. Dep. BAAS 30, fols 51–95, Typescript report of the Committee appointed to consider Reorganisation of Sections, containing original correspondence: quotes at fols 61–2.

5 MS. Dep. BAAS, fol. 66, W. N. Shaw to J. J. Thomson, 13 December 1909.

6 RGS-IBG Archives, Correspondence Block CB 1881–1910, J. Keltie to O. Howarth, 10 January 1910.

7 RGS-IBG Archives, Correspondence Block CB 1881–1910, A. J. Herbertson to J. Keltie, 11 May 1910.

8 On Mackinder's geographical views, notably his geopolitics, see G. Kearns, *Geopolitics and Empire: The Legacy of Halford Mackinder* (Oxford: Oxford University Press, 2009).

9 RGS-IBG Archives, Correspondence Block CB 1881–1910, J. Keltie to O. Howarth, 10 May 1910.

10 RGS-IBG Archives, Correspondence Block CB 1881–1910, O. Howarth to J. Keltie, 11 May 1910.

11 RGS-IBG Archives, Correspondence Block CB 1881–1910, J. Keltie to A. J. Herbertson, 21 May 1910.

12 RGS-IBG Archives, Correspondence Block CB 1881–1910, W. A. Herdman to J. Keltie, 8 June 1910; J. Keltie to W. A. Herdman, 10 June 1910.

13 RGS-IBG Archives, Correspondence Block CB 1881–1910, W. A. Herdman to J. Keltie, 11 June 1910.
14 W. H. Brock (ed.), *H. E. Armstrong and the Teaching of Science, 1880–1930* (Cambridge: Cambridge University Press, 1973).
15 MS. Dep. BAAS 30, fol. 151.
16 R. Yeo, *Science in the Public Sphere: Natural Knowledge in British Culture, 1800–1860* (Aldershot: Ashgate, 2001), p. 283.
17 Close was appointed Director General of the Ordnance Survey on 18 August 1911. For Close's life and works, see T. W. Freeman, 'Charles Frederick Arden-Close, 1865–1952', *Geographers Biobibliographical Studies* 9 (1985), pp. 1–14.
18 C. Close, 'The position of geography', *Report of the Eightieth Meeting of the British Association for the Advancement of Science* (London: John Murray, 1912), pp. 436–43, quotes at p. 436. The address was also published under this title in the *Geographical Journal* 38 (1911), pp. 404–13. Citations here are from the BAAS *Report*.
19 Close, 'The position of geography', p. 443.
20 The full list of subject specialisations was: mathematical and cartographical geography, 3 per cent of all papers published in *Geographical Journal* 1906–1910; general physical geography, 10 per cent; vulcanology and seismology, 5 per cent; glaciers, 3 per cent; hydrography (potamography and limnology), 5 per cent; oceanography, 3 per cent; meteorology and climatology; terrestrial magnetism, 3 per cent; biological geography, 1 per cent; anthropology and ethnography, 3 per cent; economic and social geography, 7 per cent; explorations, 57 per cent. As Close noted, 'The list in fact covers ground occupied by several Sections of the British Association': Close, 'The position of geography', p. 437.
21 Close, 'The position of geography', p. 441.
22 Close, 'The position of geography', pp. 443, 439.
23 MS. Dep. BAAS 422, *The Times*, 8 September 1911.
24 A. J. Herbertson, 'Geography at the British Association, Portsmouth meeting, 1911', *Geographical Journal* 38 (1911), pp. 504–14. Herbertson noted only, 'Even the presidential address, already published in last month's *Journal*, was delivered mainly to professional geographers' (p. 504).
25 RGS-IBG Archives, Correspondence Block 1881–1911, C. R. Markham to J. Keltie, 4 September 1911.
26 RGS-IBG Archives, Correspondence Block CB 1881–1911, C. Close to J. Keltie, 5 September 1911.
27 MS. Dep. BAAS 30, fols 135–6.
28 RGS-IBG Archives, Correspondence Block CB 1881–1911, H. R. Mill to J. Keltie, 23 September 1911.
29 RGS-IBG Archives, Correspondence Block CB 1881–1911, A. J. Herbertson to J. Keltie, 11 May 1910.
30 RGS-IBG Archives, Mill Collection, L.B.R. MSS 3, J. Keltie to H. R. Mill, 26 September 1911. 'Darwin' is Sir Leonard Darwin, Charles Darwin's

son and, at this point, immediate past president of the RGS (1908–1911); 'Johnston' is Sir Harold H. Johnston, colonial administrator and African explorer. It is clear from mention of the word 'Society' that Mill here refers to the RGS, but the point about this social-scientific divide is applicable also to the BAAS.
31 RGS-IBG Archives, Correspondence Block CB 1881–1911, J. Keltie to G. Lyons, 26 September 1911; W. M. Davis to J. Keltie, 25 September 1911.
32 One modern commentary to note Close's debate and its impact upon geography is R. Ellen, 'Persistence and change in the relationship between anthropology and geography', *Progress in Human Geography* 12 (1988), pp. 229–62.
33 RGS-IBG Archives, Correspondence Block CB 1911–1920, J. Keltie to O. Howarth, 27 September 1912.
34 RGS-IBG Archives, Correspondence Block CB 1911–1920, W. J. Barton to J. Keltie, 29 September 1912.
35 RGS-IBG Archives, Correspondence Block CB 1911–1920, J. Keltie to W. J. Barton, 1 October 1912.
36 On Spiers Bruce, Rudmose Brown and the Scottish National Antarctic expedition, see Peter Speak, 'Williams Spiers Bruce, 1867–1921', *Geographers Biobibliographical Studies* 17 (1997), pp. 17–25; C.W.J. Withers, *Geography, Science and National Identity: Scotland since 1520* (Cambridge: Cambridge University Press, 2001), pp. 217–9; R. N. Rudmose Brown, *A Naturalist at the Poles* (London: Seeley Service, 1923).
37 Scott Polar Research Institute, MS. 101/19/42, R. N. Rudmose Brown to W. S. Bruce, 19 April 1912.
38 Scott Polar Research Institute, MS. 356/46/16, W. S. Bruce to R. N. Rudmose Brown, 24 April 1912.
39 [Anon.], 'Geography at the British Association', *Scottish Geographical Magazine* 28 (1912), pp. 571–88, quote at p. 574.
40 [Anon.], 'Geography at the British Association' (1912), pp. 587–8.
41 S. Beaver, 'Geography in the British Association for the Advancement of Science', *Geographical Journal* 148 (1982), pp. 173–81, quote at p. 177.
42 See for example MS. Dep. BAAS 31, Papers of the Committee of the Council on the Working of the Association, 1918–1922; MS. Dep. BAAS 97, British Association Subject Files, 1920–1930; MS. Dep. BAAS 100, British Association Subject Files, 1933–1934; MS. Dep. BAAS 104, Correspondence concerning the British Science Guild, 1922–1943; MS. Dep. BAAS 107, Printed Minutes of the Conjoint Board of Scientific Societies; MS. Dep. BAAS 108, Correspondence with various Societies and Institutions, 1906–1949. Even allowing for the problems of absence of evidence and evidence of absence, the fact that these records survive suggests a step-change in the nature and purpose of the BAAS after 1919.
43 On this point, see also W. McGucken, 'The social relations of science: the British Association for the Advancement of Science, 1931–1946', *Proceedings of the American Philosophical Society* 123 (1979), pp. 237–64.

44 T. W. F. Parkinson, 'Geography in the curriculum of higher education', *Report of the Eighty-seventh Meeting of the British Association for the Advancement of Science* (London: John Murray, 1919), pp. 444–8, quote at p. 446.
45 MS. Dep. BAAS 333 [Brief notes on the National Geography Committee 1920–1936]; Royal Society Archives, CMB/103a British National Committee for Geography, 1934–1952. At the meeting of 9 April 1934 members of the committee present were Col. Sir Henry Lyons (chair), the Hydrographer of the Navy, the Director General of the Ordnance Survey, the President of the RGS, Colonel Sir Charles Close, Professor H. J. Fleure, Sir William Foster, Professor K. Mason and A. R. Hinks (Secretary to the RGS).
46 My remarks on this example owe much to P. Collins, 'The British Association for the Advancement of Science and Public Attitudes to Science, 1919–1945', unpublished PhD thesis, University of Leeds, 1978, especially pp. 306–17.
47 MS. Dep. BAAS 332, Section E, Correspondence and Papers of the Committee on the Teaching of Geography, fols 30–1, 31 March 1921.
48 MS. Dep. BAAS, fol. 117, 31 May 1922.
49 P. M. Roxby, 'The scope and aims of human geography', *Report of the Ninety-eighth Meeting of the British Association for the Advancement of Science* (London: John Murray, 1931), pp. 92–104, quote at p. 104.
50 *Liverpool Daily Post*, 21 August 1923, p. 15.
51 On Cornish's blend of eugenics and imperialism, see K.-J. Dodds, 'Eugenics, fantasies of empire and inverted Whiggism: an essay on the political geography of Vaughan Cornish', *Political Geography* 13 (1994), pp. 85–99.
52 V. Cornish (1924), 'The geographical position of the British Empire', *Report of the Ninety-first Meeting of the British Association for the Advancement of Science* (London, Murray), pp. 126–37, quotes from pp. 130 and 137 respectively.
53 Cornish, 'The geographical position of the British Empire', p. 133.
54 *Daily Herald*, 14 September 1923, in MS. Dep. BAAS 425.
55 *Daily Telegraph*, 15 September 1923, in MS. Dep. BAAS 425.
56 MS. Dep. BAAS 422, Scrapbook 3, Press cuttings, 1911–1913, *Birmingham Post*, 7 June 1913.
57 The 'handbook' in 1859 for the Cheltenham meeting was in reality a collection of papers relating to that meeting, brought together by Mrs Margaret Fison and which had appeared first in the periodical *Leisure Hour*.
58 G. F. S. Elliot, M. Laurie and M. J. Barclay (eds), *Flora, Fauna and Geology of the Clyde Area* (Glasgow: Maclehose, 1901), p. v.
59 A. W. Paton and A. H. Millar, *Handbook and Guide to Dundee and District* (Dundee: Caird, 1912), p. xiv.
60 On regional survey in Britain in this period, see D. Matless, 'Regional surveys and local knowledges: the geographical imagination in Britain, 1918–1939', *Transactions of the Institute of British Geographers* 17, (1992), pp. 464–80.

61 MS. Dep. BAAS 253, Correspondence relating to Regional Surveys, 1925–1926, fol. 1 [no date].
62 MS. Dep. BAAS 253, Correspondence relating to Regional Surveys, 1925–1926, fol. 2 [no date]. Some indication of the ways in which regional survey was to be tied to the work of the BAAS is evident in a letter from the leading geographer H. J. Fleure to John Linton Myres: 'My idea . . . was that the B.A. Correspon. Socs. Cttee. should take an interest in the places the B.A. proposes to visit & should stimulate local groups to include in their preparations for the B.A. visit the setting up of an exhibition which could be more or less permanent (i.e. the drafting of maps of social & cultural distributions which would form the nucleus of a permanent local collection to be stored and exhibited from time to time) & along with the maps there would be pictures, diagrams + so on to supplement the local museum. My feeling is that if we could press that in a few cases it would spread over the country + that our Regl. Surveys Cttee. of the G.A. [Geographical Association] could help enormously here.' MS. Dep. BAAS 253, fol. 3, Fleure to Myres, 12 November 1925.
63 MS. Dep. BAAS 253, fol. 6, Russell to Myres, 14 December 1925.
64 MS. Dep. BAAS 253, fol. 13, Sheppard to Howarth, 23 December 1925.
65 MS. Dep. BAAS 253, Myres to Howarth, fol. 14, 31 December 1925.
66 RGS-IBG Archives, Correspondence Block CB 1921–1930, J. N. L. Myres to E. A. Heawood, 9 January 1928.
67 T. W. Freeman, 'Charles Bungay Fawcett, 1883–1952', *Geographers Biobibliographical Studies* 6 (1982), pp. 39–46.
68 MS. Dep. BAAS 253, fols 7–9, C. B. Fawcett to J. N. L. Myres, 17 December 1925.
69 RGS-IBG Archives, Correspondence Block CB 1921–1930, Myres to Heawood, 9 January 1928.
70 The fact that proposals had been advanced in 1920 with a view to 'standardising them as a series, and making them available to the public' suggests a longer gestation to the model finally realised in the *Scientific Survey* volumes from 1932: MS. Dep. BAAS, Council Minutes, 1920–1929, fol. 33, 10–12 April 1920. To judge from contemporary reviews, the move to such survey volumes was welcomed because it established a standard format 'on a definite plan, in lieu of the former unsystematised handbooks': *Nature*, 19 August 1933.
71 A. Stevens, 'Address to section E', *British Association for the Advancement of Science Dundee Meeting* (London: British Association, 1939), p. 1.
72 On Newbigin's work and as a pioneer woman geographer, see A. Maddrell, *Complex Locations: Women's Geographical Work in the UK, 1850–1870* (London: Wiley-Blackwell, 2009), pp. 60–77.
73 RGS-IBG Archives, Correspondence Block CB 1921–1930, Letter of A. R. Hinks, 22 November 1929. Hinks directed his letter, on Fawcett's recommendation, to 'Fleure, Miss Beatrice Hosgood [he meant Blanche Hosgood, Reader in Geography at Bedford College, London, from 1923 to 1948],

Rodwell Jones, Roxby, Unstead, Debenham, Brigadier Jack, Dr. Longstaff, Mackinder, Sir Matthew Nathan, Mr. Wollaston, and Mr Wordie'.
74 RGS-IBG Archives, Correspondence Block CB 1921–1930, Human Geography Committee [summary of committee dates].
75 RGS-IBG Archives, Correspondence Block CB 1921–1930, F. Debenham to A. R. Hinks, 25 November 1929.
76 H. R. Mill, *Life Interests of a Geographer: An Experiment in Autobiography* (East Grinstead: published privately, 1945), p. 86.
77 University of Cambridge, Scott Polar Research Institute, MS. 1418/3/1–8, 21 January 1935.
78 [Professor H. E. Armstrong in *The Chemical Age*, 7 October 1933]: MS. Dep. BAAS 427.

8

Conclusion: the British Association, geographies of science and the science of geography

'Parliament of Science' – even 'Imperial Parliament of Science'; parasitic 'little commonwealth'; 'British Association for the Advancement of Everything in General and Nothing in Particular'; 'gigantic boa constrictor'; 'Mudfog Association for the Advancement of Everything'. These and other names have been used at one time or another to label the British Association for the Advancement of Science. For MacLeod, considering the shift away from conformity with established religion towards the new 'priestcraft' of science in the later nineteenth century, together with that body's intrinsically peripatetic nature, 'The British Association may have seemed less the Parliament of Science than its travelling seminary.'[1]

In examining the British Association from its foundation in 1831 to the events of the 1930s I have been concerned less with what it was called than with where and how and why it was at work and with the implications of its geographical mobility for the advancement of science. For the case of geography as one of its constituent sciences, I have been concerned with where and how and why that science was undertaken, by whom, for whom, and with what consequence for geography's different audiences. This twin focus upon the geography of science and the science of geography in historical context is grounded in that rich array of work undertaken by historians of science and historians of geography which, as for the BAAS, has in turn enjoyed various descriptors – the 'spatialised historiography of science', the 'spatial turn' in science studies, 'geographies of science'. It is intended, of course, as a contribution to that wider endeavour as, too, it is a contribution to the more specific historiography relating to the British Association.

In closing, I want to review the principal findings and interpretations relating to the two main issues that have structured this enquiry – the geography of BAAS science and the science of BAAS geography – and, in so doing, consider some of the wider implications as they relate to geographical work in the history of science on the one hand and to the history of British geography on the other. Such questions are closely related. If this account has shown anything, it is that geography emphatically was seen as a science as that term emerged in and from the 1830s in civic context in Britain and was given institutional shape and different content as a professional pursuit, a suite of methodological procedures and as a socially defined subject matter, a 'discipline', a 'field'. Yet sectional and institutional identity – for geography, as section E, from 1851 at least – was not the same thing as epistemological 'definition', a bespoke focus that, in the eyes of practitioners and audiences, delimited one science from another, however broadly. As I have also shown, geography was by some people (geographers included) and for different reasons equally emphatically reckoned 'unscientific', not worthy of the status contemporaries accorded the term 'science'.

To disclose in these ways elements of the geography of British science in civic and geographical context in the century from 1831 and of geography's presence in the BAAS – for some a popular civic practice, for others an emergent body of scientific knowledge, for yet others not a science at all – suggests that the study offers some insight into our continuing enquiries concerning science's making in geographical context and the history in civic context of geography as subject and science.

Historical geographies of BAAS science

The decision at foundation 'ever to be Provincials' had significant consequences for the BAAS and for science's audiences throughout Britain and Ireland and the Dominions. Geographical mobility was an operational strategy designed to bring science to the public. In this sense of geography as location and strategy, the BAAS was, so to speak, geographically minded from the off. The dissemination of science – if it was not to be the preserve of metropolitan elites and of their audiences – required geographical mobility. In the absence of a periodical which could transmit news of BAAS affairs or the findings of its practitioners, what made BAAS science move was the fact that its practitioners and publics did. Newspaper reports carried news of these men of science and their audiences to others who had not seen or heard things for themselves.

After an initial period, 1831–c. 1845, in which the circulation of BAAS annual meetings was designed to secure the social and institutional status of leading academic participants, BAAS meetings were from the mid nineteenth century held in towns where a culture of science and learning was already active, where extant civic bodies for the promotion of natural and social knowledge co-ordinated an invitation and or where the BAAS had not hitherto been. In deciding amongst competing invitations where the Association would meet two years hence, Council on more than one occasion prioritised the town not before visited: science needed to have, and to be seen to have, national coverage. Visits required institutional orchestration and civic deputation. Return visits were more likely to happen if the town as a host venue offered the physical and intellectual 'fabric' necessary for scientific enquiry – citizens of repute in some branch of learning, for example, in the form of a university or college, learned societies, museums, even a town hall of sufficient size – or if the town or city had developed sufficiently over time in ways deemed likely to attract delegates. Meetings could be delayed or not run smoothly if civic space was insufficient (as in the case of Leeds and its town hall in the 1850s), when sectional meetings were held too far apart from one another (as geographers found in Glasgow in 1901), or, as was reported on many occasions and usually of presidential addresses and papers by leading scientific celebrities, if meeting rooms were too cramped even to see or hear the distinguished scientist. In these senses, BAAS meetings and the public's appreciation of its different sciences were conditioned by the location, the physical character of the venue.

This is, nevertheless, a relatively weak sense of the geography of BAAS science – of an urban geography as science's physical 'container', albeit one at work at different scales: from the whole town or city, to the urban circuitry required of participants and delegates between sites of active attendance and passive 'looking on', to the nature of individual buildings or rooms. But BAAS meetings were not, precisely, urban affairs. I mean both that the whole town was not usually a site for science – only certain civic knowledge spaces were – and that the local area was often utilised to illustrate certain aspects of scientific work: for botanical fieldwork, or geological excursions, as an illustration of a 'type site' which would, for the duration of the meeting, be a 'field sight'. Documenting such sites/sights in the form of essays in the meeting handbooks, for example, was a means of securing in print a reputation as 'local expert' – a term that I think more apt than 'professional' or 'amateur' in relation to the BAAS – and of contributing to the exchange of knowledge to visitors and those beyond local borders by highlighting specific features in an area's regional geography.

There is, however, a stronger social sense to science's geography. I mean, in part, that the geography of a town or area could and many times did determine what was in the programme for BAAS annual meetings. Papers on the chemistry of local minerals; on the dependence of the local economy upon imperial commerce (be that Manchester's industrial goods in 1861, Winnipeg's wheat in 1909 or Liverpool's export of people in 1923); even on the symbolic importance of engineering to a town's civic identity as in Sydney and the programme for section G in 1914: each of these and many others demonstrate that the content of BAAS science was, to varying degrees and in different ways, always shaped by local interests. The fact that in a period of mounting public anxiety about the nature, benefit and future importance of science BAAS central committees sought in 1910 to re-emphasise that science should speak to local interests is noteworthy in this respect.

I mean also we should not see BAAS science as something simply imported as a 'finished' product into its venues – nor, of course, that we should see it as simply reflecting the needs of local audiences there. We should, rather, see it as something made and reacted to in place. Place is not a location, 'mere background atmospherics', as one author terms it, but a social construction which, in providing for the requirements of local 'scientists', resident audiences, visiting speakers and the opportunities provided by a town and its setting, established different possibilities for intellectual debate.[2]

This interpretation of science as always social and always locally conditioned does not rest upon a series of localist city-based studies. The city is not the required unit of scale for examining the geography of BAAS science in urban context. And its utility for an understanding of science's geographies is not diminished in incorporating the imperial and international agenda of the BAAS. Here, too, the BAAS was made to be mobile, its officers promoting an imperial agenda, its delegates using the fact of their international travel to meetings to do science on the way. Once overseas, BAAS delegates took part in meetings which were themselves mobile: long-distance rail travel was required to see the realms visiting scientists had to know in order to benefit them as, in turn, host communities and type sites/sights displayed their scientific credentials. Organisers of the BAAS overseas meetings were just as concerned as their British and Irish counterparts, if not more so, that their meetings should reflect and direct the needs of locals and that they should balance local matters within sections' programmes with work of general or thematic significance. This view of the geographies of BAAS science as not just 'contained' in locational settings but socially constituted in place and by circulation across national and international space is reinforced by what we have seen of the ways in which the annual meetings worked.

BAAS meetings were locatable and datable scientific events but they were not unified events of science alone. This study has suggested that they be better regarded as situated moments of practice, as different sited forms of communicative action which, given their different sites and forms, attracted not different audiences so much as audiences of differing purpose according to which subject they went to, why they went where they did – to hear or see the scientists, be seen as part of the meeting's 'social set', attend a microscopical *soirée*, and so on. As Knight notes, BAAS meetings 'were an excellent opportunity to take an intellectually stimulating holiday, being entertained by guest lecturers, meeting old friends and seeing local sights on guided tours'.[3] Admittedly, the focus has been on geography and on some practices more than others – on presidential addresses, sectional programmes, excursions, the production and use of handbooks to the relative neglect of social events. And, for geography, I have looked also at the funding of its grants and the wording of its recommendations by which its content and direction was defined and distinguished in its sectional identity in relation to other emergent sciences. Experiencing geography was, however, quite different from managing its sectional purpose and from other subjects: from the relatively rigorous demands of section A, for example, or from the at times confusing conceptual terminology of section C, even if, with delegates to sections C and D and K, field excursions were an experience shared by many. Experiencing geography as a BAAS science differed between its professed and professional practitioners and its professional and populist audiences and did so for reasons to do with competing notions as to the subject's content, defining methods and purpose.

In addressing questions about the social spaces of scientific practice, it is appropriate to recognise that whilst the BAAS saw and sustained itself as an institution embodying expertise through its associated members, it worked in practice in public in civil society, 'instituting' and 'inscribing' science as Lenoir has it, in quite different ways.[4] I mean, simply, that the lecture platform was not the laboratory or the open spaces encountered on expeditions or excursions. The botanical or geological textual outlines in BAAS handbooks guided one in the field, but taxonomic classification often took place away from sites of collection. Of those attending science's 'speech sites', how many followed the science in its professional published outlets, or charted later developments over time in how a given science functioned? For practitioners in the meetings, BAAS spoken papers were not so much science work, science at work, so to speak, as they were declarations about science's work, statements about that science's epistemological reach or its public accountability, the science itself having been done elsewhere, even by others. For the recipient, on the other hand, what lay as science 'behind' the performance of science was, perhaps, unimportant if the intention in going was to be educated and informed without feeling overwhelmed by what was being said or how. This is not to place what others have described as scientists' laboratory 'shop talk' or the language of specialisation, written or spoken, as in intellectual and social terms somehow above the language of the podium: public speaking is no less a scientific skill than authorship.[5] It is to emphasise of the BAAS meetings that science as spoken and listened to was more crucial than science as written or read, that at least an element of credibility attached to the speaker's perceived celebrity in the public's eyes, and that, for some in the audience, attending was a social act first and a scientific one second: Helen Shipton referred to a woman at the 1892 Edinburgh meeting who declared herself there only for the purpose of '"Looking on!" or studying character – a study for which there is here a wide and interesting field'.[6]

The evidence suggests no clear or single demarcation to be possible in categorising the audiences for BAAS science. The observation at foundation that there would be two different audiences, 'two interested classes' – 'visitors coming to meet their fellows in science and the intelligent persons of our own neighbourhoods who hope to be gratified and instructed' (see p. 28) – seems to have been borne out in general terms for the following decades. There is evidence, however, of a shift over time in the term 'fellows in science' between professed and professional men, and latterly women, of science. From the later nineteenth century onwards, disciplinary formation and institutionalisation outside the BAAS meant that, increasingly, its officers and its experts were drawn from subjects' professional ranks. Speakers in sectional programmes could be a mix of specialists (often self-professed and without relevant institutional credibility who need not be professionals) and professional men. Together, such individuals illustrate differences within what we might think of as the 'subject-informed' active participants within BAAS meetings. This study has also discerned differences in general terms within the mass of 'intelligent persons', whether from the locality or from farther afield: between men and women, in relation to given sections, and, for the case of geography, in relation to types of paper within section E.

Taken together, then, we can think of the geographical dimensions of the BAAS as evident in several ways. The operation of its mission as an association for the promotion of science as a cultural good depended upon geographical mobility. This was evident as an operational stratagem at foundation at local and national scales, and, from 1884, at the transnational scale as the BAAS helped realise a colonial agenda for science in the British dominions. The principal means employed by the BAAS for promoting science as a civic project was its annual meeting. BAAS meetings were shaped by their location, by the conjunction of social and scientific actors in a place and by the geography of the local area wherein certain features would be used, almost as scientific credentials, to attract visitors to BAAS meetings. Within the city, certain venues of practice – principally, sites of speech but also sites of display – acted as sites for science's 'making', public dissemination and, simultaneously, of science's reception. The intention and comprehension of audiences have not proven to be easily recoverable. But their constitution and comprehension varied between towns and within civic venues according to scientific subject, the emphasis given to papers about or relating to the local area, the type of paper and its public performance within the science in question, and, on occasion, the status of the speaker.[7]

This explanation of BAAS science as geographically constituted through institutional movement, in location and as situated social practice does not deny the legitimacy of terms such as 'production', 'reception' and 'mobility' in the conceptualisation of science's geographies. But it does begin to break them down, to refine them in relation to the making and reception of science in place and over space. It is to suggest, too, that notions of 'site' or 'setting', 'scale' and 'practice' have perhaps greater explanatory potential in revealing how the BAAS worked where it did. This has been shown to be constituted as local geographies of practice and of differential comprehension in particular civic settings, whether in overseas meetings, at the level of the nation or in relation to particular urban geographies.

The science of BAAS geography

In considering the nature and place of geography as a science within the BAAS, we should not lose sight of the fact that as a subject embracing the description of the whole earth geography was well established before 1831, in Britain and elsewhere, that it had long been taught in Oxford, Cambridge and in several of the Scottish universities and that, in terms of a publishing tradition, books of geography were numerous and had been widely used in education since the late seventeenth century.[8] So, too, of course, those other subjects identified as sciences within the BAAS and shaped by it had a prehistory before 1831. The term 'science' also had prior currency as a branch of learning and as the methods by which observed facts systematically classified were the bases to general laws explaining nature.[9] Given this, we may in limited ways identify geography as a 'science' before 1831.

Geography was a science in the BAAS in different ways. Its officers and others institutionally 'defined' it as such. It took shape in terms of questions of natural and phenomenal distribution over space; in subcommittee identity from 1831 to 1833; shared sectional identity with geology from 1833 to 1851 as subjects interrogating and mapping the earth's physical dimensions, and, from 1851, as a sectional identity shared with ethnology until 1871. At the same time, however, there were in 1831 no men who, as 'geographer', might qualify for the term 'scientist'. This was not because the latter term, strictly, had not yet been invented by Whewell, but stems from the fact that the former term did not enjoy equivalent status. If it had currency at all, the term 'geographer' spoke more to textual hackwork by journalists than adherence to scientific methodology by natural philosophers.[10] What was covered by geography as the practice of discovery and earth description embodied in the map and undertaken by naval men, via expeditions and through direct encounter was embraced in a British context by the term 'explorer', *voyageurs-naturaliste* in French, and was undertaken generally as a practice of primary survey, not of law-based mathematical enquiry. There are, perhaps, parallels with other BAAS sciences. For Endersby, writing of natural history work within the BAAS, 'the pattern is too strong to ignore – not only was natural history poorly funded, but within Section D botany was clearly zoology's poor relation'.[11] Geography was, arguably, geology's poor relation with respect to the physical earth sciences and, after 1851, in embracing the nascent human sciences, only added to its imprecision in adding the uncertainty of human distribution to its sectional make-up.

We may thus make distinctions after 1831 between geography as sectional and institutional subject, geography as a discipline in terms of the means by which it worked to carve out a certain intellectual territory, and geography in terms of those practices by which it disciplined itself and was seen to have a certain content and purpose by different audiences. The connections overall between subject, discipline and practice show geography's making in civic context as a 'field', taking this last term to mean the 'positions occupied by agents with differential stances toward one another. The field is not just a domain of differentially constructed intellectual positions; it also includes instruments of circulation', has shifting boundaries with other fields (and with what was and was not science) and was shaped in different places.[12]

Geography's early sectional history within the BAAS and, particularly, its conjoint sectional identity – with geology in one period (1833–51) and ethnology in another (1851–71) – reflected the uncertainty of science's emergence and management within the BAAS more than it did any epistemological imprecision as to what geography was. This is clear with respect to that understanding at the outset that geography's focus lay with spatial distribution and only latterly with causal explanation of that distribution, of the subject's connections with geology and with botany and of the emphasis upon exploration and in mapping. But it is true to say that the science of geography embraced questions to do with the description, distribution and explanation of physical phenomena earlier and more easily than it did with human phenomena. Human subjects did not so readily present themselves as sectional subjects. Human geography emerged as a disciplinary concern within the field of geography much later than the geography of humans appeared as a subject concern for sections E and H. Sections worked within the BAAS to identify certain subject areas and to facilitate access to the limited funds to promote research. The geographers in section E associated with members of virtually every other section at one time or another, but did so chiefly with geologists and botanists, with agriculture and statistics latterly, and, in individual meetings and for given purposes, came together with the mathematicians and physicists over geodetic survey or, in the 1920s particularly, with education in addressing the common agenda of citizenship and empire.

The focus on exploration, mapping and survey meant that geography's identity and its disciplining procedures within section E were shaped by close connections with the Royal Geographical Society in particular and by the emphasis that body accorded exploration, expeditionary endeavour and imperial promotion. This institutional-epistemological affiliation was a reflection of the shared membership of these bodies, the role of influential figures such as Roderick Murchison, Clements Markham, J. S. Keltie and H. R. Mill and of the institutionalisation of the physical sciences more generally in the nineteenth century. But these individual, institutional and purposive connections did not mean that geography as a section within the BAAS was at all times served by professional geographers. Rather, as within the RGS, geography was from the 1830s until the later 1880s promoted and practised in the BAAS by men I have here termed 'professed geographers' more than by 'professional geographers', using the first term to identify men whose interests lay in earth description, exploration and phenomenal distribution and the latter those persons employed by an institution to undertake the pursuit of disciplinary knowledge under that identifying subject label.

From the later 1880s – Keltie's remarks in 1897 about a 'school of young geographers' are illuminative – there was a scholarly community in universities that we may term 'geography's professionals'. Many individuals within this community were involved with the BAAS. Given persistent emphases within the RGS over exploratory work rather more than thematic research, especially in human geography, they saw section E as the means to promote more specialist work. At the same time, and certainly by the second decade of the twentieth century, the work of BAAS section E was being hindered by the refusal of some academic geographers in the subject's epistemic community to participate fully in its activities even as others saw it as a vital forum for geography's promotion as a discipline and for public engagement with geography. Other fault lines were by then evident within BAAS section E and in British geography. Audiences identified a lack of central purpose even as they continued to be attracted to popular talks. Physical geographers saw little value in the case studies of human geographers even as the BAAS in the mid 1920s used regional geography as the methodological basis to its handbooks, a means to reaffirm local identity and the scientific credentials of the locations chosen for its meetings. Seemingly shared concerns in the 1920s over geography and citizenship embraced divisions between those with a global vision and those with a British imperial vision for geography's educational role.

The most distinctive feature of geography's place as a BAAS science – apparent particularly from the mid 1870s – was the separation between 'popular geography' and 'scientific geography'. For the former, its exploratory and expository traditions were rendered first-hand by men whose celebrity status as 'geographer' was usually attained by fieldwork in adversarial environments and ascribed by audiences who relished accounts of the far-away and hazardous. For the latter, geography's increasing scientific emphasis was evident in claims by men of science about more precise methods, thematic not general work and the application of geographical thinking to causal explanation of natural and social phenomena. Geography was made 'real' for its audiences through spoken papers, through oratorical performance, often accompanied by the use of maps or lantern slides, more than through journal articles. Hearing and seeing leading explorers about whom one had read was a key component of the nature and popularity of section E. That fact, and the fact that narratives of exploration and other general papers of a descriptive nature required little prior knowledge, meant that section E attracted a high proportion of people with no scientific education, and many women. Claims from the mid 1870s that geography should be characterised either by a different sort of exploration or – and this from the 1890s onwards – by more specialist work and via more evidently scientific procedures were made most commonly in presidential addresses. The incorporation of local themes within the general programme of papers in section E was a common practice, albeit that it had to be reinforced at moments in order to ensure a connection between the nature of the meeting and the interests of host audiences.

Divisions over geography's scientific status from within the scientific community were sharply apparent from the second decade of the twentieth century: notably in Close's 1911 Portsmouth address, but also in the views of some geographers and of scientists in other sections, and in the 1920s they were evident in debates over geography's scientific status in terms of educational funding. Reservations hinged upon geography's epistemological focus, or, more precisely, its lack of one. The view was held that physical geography might embrace the deterministic causal principles of sciences such as geology (a view which underlay those plans in 1910 to reunite the two in a section). Yet human geography's perceived lack of focus and a still evident exploratory tinge to the activities of section E meant that geography appeared to its detractors to be just a synthesis of other sciences' data and methods, distinguished only by its attention to mapping as a form of territorial representation and political governance.

In thinking about BAAS geography in its public and civic context and its presence and purpose as both popular and scientific but for different people, one difficulty lies in knowing how to place geography's variant practices and definitions and the work of section E in the category 'popular science', or even whether that category is meaningful. This may in turn reflect the need as Topham puts it, to rethink 'the distinction between the making and the communication of knowledge that has so bedevilled the historiography of popular science'.[13] Part of what at various times was held to be geography by some practitioners and a large proportion of the audiences was popular but others did not think it science. What, later, was deemed more scientific was less popular. The work undertaken by geographers, professed and professional, within section E was part of the promotion of science in civic context even though, at moments, other scientists and other geographers doubted geography's scientific credentials. At the same time, what passed as geography in BAAS section E was connected with the emergence of the discipline and of the subject in the universities and in other specialist institutions, but really so only from the late nineteenth century. And as geography was established within educational institutions some professional practitioners thought BAAS section E still too popular in purpose even as others recalled with affection its role in advancing the science.

Historical geographies of science and of geography

This study has revealed – more, perhaps, than for many studies of nineteenth-century science in civic context and certainly more than for existing studies of geography in historical context – the nature and range of contemporaries' claims over what science was, what geography was as a science or not and who the audiences were for those conceptions. Geography's practitioners and audiences in the century from 1831 have been brought into sharper focus and revealed in different civic contexts.

The various practices of communicative action by which science and geography were undertaken in the BAAS have been more clearly defined. Geography's institutional, sectional and epistemological relations with other sciences and practices have been examined.

It is, however, quite another thing to presume – I hope I have not – that this historical study of nineteenth-century British geography as a science within one admittedly formative institution straightforwardly stands for things beyond itself. Of course, it offers a corrective in general terms to studies of disciplinary formation in the nineteenth century which have neglected geography as a science, just as, in paying attention to one institution and its attendant practises, it complicates others' conceptions of the chronology of public science in nineteenth-century Britain.[14] This is, however, a study of only one association for the public promotion of science and of one science in particular in association with others: it is not a historical geography of all of British science in its civic and institutional context in the century from 1831 or even of science's political settings in this period. This is, with others, a study in knowing how and why one form of science, geography, and one type of BAAS 'scientist', explorer-geographers such as David Livingstone, Henry Morton Stanley and Fridtjof Nansen, should be so popular as to attract crowds in their thousands yet, simultaneously, be the subject of disdain from their fellows: it is, then, one contribution to continuing enquiries into the forms and the geographies of 'popular science' as an historical and analytic category.[15]

The book has shown for one institution and through assessment of its operating practices how the business of that body, the nature of its science, was shaped by the local settings in which its meetings met. It has shown how geography was put to shape as a civic science outside the confines of educational institutions yet, often, was made to work in concert with them over imperial and international concerns for science as a project of education for Britain's citizens, at home and abroad. This has not been a study of disciplinary formation or professionalisation, of one science's emergence as a professional 'field'. It should not be read teleologically as a historical precursor to geography's 'modernity' from the later nineteenth century for the good reason that it decisively challenges the terminology and the chronology and any conceptualisation of geography and science's simple essentialist development over time and across space. It adds to our knowledge of geography's history in its other institutional, mainly academic, settings but differs from them and from others' focus on the making of geographical knowledge. Introducing their *Nineteenth-Century Geographies*, for example, Michie and Thomas began thus:

The nineteenth century was a time of unprecedented discovery and exploration throughout the globe, a period when the 'blank spaces on the earth' were systematically investigated, occupied, and exploited by the major imperial powers of Western Europe and the United States. Expeditions were launched in search of the North West Passage, treks to the North and South Poles were dispatched, the Indian subcontinent was surveyed, the sources of the great rivers of Africa were tracked down and mapped, the Far East was opened to Western markets, and Mount Everest was discovered by Western explorers.[16]

My attention to the workings of BAAS section E has shown how and by whom such 'grand scale' geography was undertaken, has looked at such geography's reception in civic context and at the mechanisms by which geography was locally put to work.

These facts have wider purchase. One implication may be, as others have suggested, to disrupt the conventional 'cartographies' of science's geography:

> To write a historical geography of science is explicitly to present scientific knowledge and practice as phenomena that are made in place and shaped by spatial relationships. Traditional parameters of location, distance and the journeys in between must be at the core of any such project, but it is important to include those aspects of space and place, as understood in contemporary geographical thought, that confound cartography. A full history of science requires us to extend summaries of location, distance and journey to create a new register of positionings, spacings and traffickings.[17]

What has been presented here is elements of a possible 'new register': certainly so in terms of the civic 'spacings' or 'settings' in which the BAAS worked, in and out of town, and of the meetings' 'traffickings' as people moved to excursions and to hear lectures. This is not, however, to give the BAAS some sort of institutional agency in absolute terms. Rather, it is to stress the different material and communicative means, the lecture talk, the presidential address as 'disciplinary' pronouncement (and denouncement), the excursion, mobility and international travel as key practices by which geography as a BAAS science was made and communicated. In looking at how it worked, the BAAS as one 'thing' has been lost sight of: in its stead, different practices are revealed in different venues.

It may be possible, of course, to look comparatively at science's practices within the BAAS or within other institutions in different ways and in different places. This book may be considered a study in the historical geography of BAAS geography as a civic science. What has been revealed for geography as a civic science cannot be presumed to hold for (say) botany, with which geography shared much, or for mathematics, with which it shared least in terms of explanatory focus as a science. But since the BAAS operated in such ways for all sections, and sections functioned to common purpose, there is no reason to suppose that the practical management of geography as a section was different from any other. For these or other possibilities or for science in other periods, the function and purpose of science as a form of knowledge and of cultural authority is always likely to be inflected in one way or another by the places in which it was made, communicated and received. Science has long had its histories. So, too, science has its geographies. The task in further studies of the historical geographies of science is to reveal the differences that geographical questions make to explanation of the nature of science and to science's conduct, practitioners and audiences.

Notes

1 R. MacLeod, 'Introduction. On the advancement of science', in R. MacLeod and P. Collins (eds), *The Parliament of Science: The British Association for the Advancement of Science, 1831–1981* (Northwood: Science Reviews, 1981), pp. 17–42, quote at p. 31.
2 T. Barnes, 'Placing ideas: *genus loci*, heterotopia, and geography's quantitative revolution', *Progress in Human Geography* 28 (2004), pp. 1–31, quote at p. 4. The social production of space is stressed in H. Lefebvre, *The Production of Space*, translated by D. Nicholson-Smith (Oxford: Blackwell, 1991).
3 D. Knight, *Public Understanding of Science: A History of Communicating Scientific Ideas* (London and New York: Routledge, 2006), p. 86.
4 T. Lenoir, *Instituting Science: The Cultural Production of Scientific Disciplines* (Stanford CA: Stanford University Press, 1997); T. Lenoir (ed.), *Inscribing Science: Scientific Texts and the Materiality of Communication* (Stanford CA: Stanford University Press, 1998).
5 On these points, see S. Collini, *Absent Minds: Intellectuals in Britain* (Oxford: Oxford University Press, 2006); J. Secord, 'How scientific conversation became shop talk', *Transactions of the Royal Historical Society* 17 (2007), pp. 129–56.
6 R. Higgitt and C. W. J. Withers, 'Science and sociability: women as audience at the British Association for the Advancement of Science, 1831–1901', *Isis* 99 (2008), pp. 1–27, quote at p. 21.
7 For one visitor, Rita Harris Dutton, who attended the Bristol meeting in 1930 and principally section K, Botany, her experience of the event was coloured less by the science than by the fact that at previous meetings she had attended (Leeds in 1927, Liverpool in 1923) the BAAS had paid tram fares, but not in Bristol – 'the North is more generous than the south': Lancashire Record Office, DDY 1127/1/3, Diary of Rita Harris Dutton.

8 D. N. Livingstone, 'British geography, 1500–1900: an imprecise review', in R. Johnston and M. Williams (eds), *A Century of British Geography* (Oxford: Oxford University Press in association with the British Academy, 2003), pp. 11–44; R. Mayhew, 'Geography in eighteenth-century British education', *Paedagogica Historica* 34 (1998), pp. 731–69; C. W. J. Withers and R. Mayhew, 'Re-thinking "disciplinary" history: geography in British universities, c. 1580–1887', *Transactions of the Institute of British Geographers* 27 (2002), pp. 11–29.
9 See the essays in R. Porter (ed.), *The Cambridge History of Science,* Vol. 4, *Eighteenth-Century Science* (Cambridge: Cambridge University Press, 2003).
10 C. W. J. Withers, *Placing the Enlightenment: Thinking Geographically about the Age of Reason* (Chicago: University of Chicago Press, 2007), p. 179.
11 J. Endersby, *Imperial Nature: Joseph Hooker and the Practices of Victorian Science* (Chicago: University of Chicago Press, 2008), p. 37.
12 Lenoir, *Instituting Science*, p. 15. This point about the social reconfiguration of the sciences is a central element in Cahan's overview of the making of nineteenth-century science in terms of periods, disciplines and practice: D. Cahan, 'Looking at nineteenth-century science: an introduction', in D. Cahan (ed.), *From Natural Philosophy to the Sciences: Writing the History of Nineteenth-Century Science* (Chicago: University of Chicago Press, 2003), pp. 3–15.
13 J. R. Topham, 'Introduction', *Isis* 100 (2009), pp. 310–18, quote on p. 317.
14 Cahan, *From Natural Philosophy to the Sciences,* omits discussion of geography from his overview. In discussing the nature of public science in nineteenth-century Britain, Turner notes three periods: from 1800 to 1851, in which the BAAS figured significantly in urging the importance of science as a mode of useful knowledge; the early 1850s to the early 1870s, in which science decisively challenged religion as a social force; and the 1870s to c. 1900, in which period science became strongly politicised and managed as a political resource: F. Turner, 'Public science in Britain, 1880–1919', *Isis* 71 (1980), pp. 598–608.
15 For further study of the diverse ways in which 'popular science' worked, where and how, see Topham, 'Introduction', and the papers which follow his summary in the Focus section on 'Historicizing "Popular Science"': A. W. Daum, 'Varieties of popular science and the transformations of public knowledge: some historical reflection', *Isis* 100 (2009), pp. 319–32; R. O'Connor, 'Reflections on popular science in Britain: genres, categories, and historians', *Isis* 100 (2009), pp. 333–45; K. Pandora, 'Popular science in national and transnational perspective: suggestions from the American context', *Isis* 100 (2009), pp. 346–58; B. Bensaude-Vincent, 'A historical perspective on science and its "others"', *Isis* 100 (2009), pp. 359–68.
16 H. Michie and R. R. Thomas, 'Introduction', in H. Michie and R. R. Thomas (eds), *Nineteenth-Century Geographies: The Transformation of*

Space from the Victorian Age to the American Century (New Brunswick NJ: Rutgers University Press, 2003), pp. 1–22, quote at p. 1.
17 H. Lorimer and N. Spedding, 'Locating field science: a geographical family expedition to Glen Roy, Scotland', *British Journal for the History of Science*, 38 (2005), pp. 13–34, quote at p. 33.

Bibliography

Primary unpublished sources

Adolph Basser Library, Australian Academy of Sciences
A. D. C. Rivett papers

American Geographical Society
Archives of the Association of American Geographers, W. M. Davis Part I, Box 156

Birmingham University Library
MS. Ref 4/i/3

Bodleian Library, Oxford
British Association for the Advancement of Science MSS Papers [individually referenced in chapter notes under 'MS. Dep. BAAS' and file number]
MS. Eng.e.386 Diary of Agnes Hudson

Bristol Record Office
MS. 32079

British Library
Murchison Papers, Add. MS. 46125

Cambridge University Library
MS. CUR 111.2

Capetown Provincial Archives
3/CT 4/4/1/31 HM 217/4

City of Sydney Archives
File 1910–1956 [BAAS Local Committee papers, 1914 visit]

Bibliography   249

Lancashire Record Office
DDY 1127/1/3, Diary of Rita Harris Dutton

Leeds University Library
Papers of Leeds Geological Association
Papers of Leeds Philosophical and Literary Society

Mitchell Library, Glasgow
MSS and Glasgow Corporation papers relating to the visits of the BAAS to Glasgow [1840, 1855, 1876, 1901, 1928]

National Archives of Australia
CP103/11

National Archives Repository, Pretoria
MS. GG 208 3/4902

National Library of Ireland
MS. 4792 Wicklow Papers

Newcastle upon Tyne University Library
Papers of the Literary and Philosophical Society of Newcastle upon Tyne

Oxford University, School of Geography and Environment
Mackinder papers MP/C/100

Pietermaritzburg Archives Repository
CSO 1780
SNA I/1.323

Royal Anthropological Institute, London
A1, Minutes of the Ethnological Society 1844–1869
A3, Minutes of the Anthropological Society 1863–1870
A8, Correspondence files

Royal Geographical Society (with the Institute of British Geographers)
Correspondence Block CB 1871–1880
Correspondence Block CB 1881–1910
Correspondence Block 1882–1902
Correspondence Block 1921–1930
Markham Collection [CRM]
Mill Collection [L.B.R. MSS 3]
SSC/11 File 2, Minute Book of Section E, 1865–1873

Royal Society Archives
CMB/103a British National Committee for Geography 1934–1952

Scott Polar Research Institute, University of Cambridge
MS. 1418/3/1–8
MS. 101/19/42
MS. 356/46/16

University of Toronto Archives
MS. B65-0021

Western Cape Provincial Archives and Records Services
MS. 3/CT 4/4/1/31 HM 217/4

Wits University Manuscript Collections
MS. AF 1211
MS. A. 1753 Diary of Gertrude Mabel Rose

Unpublished research theses
Collins, P. (1978) 'The British Association for the Advancement of Science and Public Attitudes to Science 1919–1945', unpublished PhD thesis, University of Leeds.

Primary published sources

Books, pamphlets, parliamentary reports
[Anon.] (1857) *A Short Description of the Western Islands of Aran, County of Galway, chiefly extracted from the Programme of the Ethnological Excursion of the British Association to these interesting Islands in the Autumn of 1857, under the direction of W. R. Wilde, M.R.I.A.* (Dublin: no publisher).
[British Association for the Advancement of Science] (1899) *Ethnological Survey of Canada* (London: John Murray).
[British Association for the Advancement of Science] (1930) *The Human Geography of Tropical Africa: The Need for Investigation* (London: British Association for the Advancement of Science).
Committee of the British Association for the Advancement of Science (1839) *Queries Respecting the Human Race, to be addressed to Travellers and Others* (London: Richard and John Taylor).
Committee on North-Western Tribes for the Dominion of Canada (1884) (Montreal: British Association for the Advancement of Science).
Frost, C. (1853) *On the Prospective Advantages of a Visit to the Town of Hull by the British Association for the Advancement of Science* (Hull: privately printed).
Haverty, M. (1857) *The Aran Isles; or, A Report of the Excursion of the Ethnological Section of the British Association from Dublin to the Western Islands of Aran in September, 1857* (Dublin: no publisher).

Memorial of the British Association for the Advancement of Science, in relation to the Present State of Survey of Great Britain, British Parliamentary Papers 1836 XLVII, pp. 89–96.

Report from the Select Committee appointed to consider the Ordnance Survey of Scotland; together with the Proceedings of the Committee, Minutes of Evidence, Appendix and Index, British Parliamentary Papers XIV (1856), pp. 75–80.

Report of the Meeting of the British Association for the Advancement of Science (London: John Murray, 1831–).

Journal articles

Aberdare, Lord (1888) 'On the promotion of the study of geography', *Report of the Fifty-seventh Meeting of the British Association for the Advancement of Science* (London: John Murray), pp. 158–9.

[Anon.] (1836) 'Resolutions of the committee on magnetism', *Report of the Fifth Meeting of the British Association for the Advancement of Science* (London: John Murray, 1835), pp. 1–34.

[Anon.] (1865) 'Anthropology and the British Association', *Anthropological Review* 3, pp. 354–71.

Baker, S. (1868) 'Geography and ethnology', *Report of the Thirty-seventh Meeting of the British Association for the Advancement of Science* (London: John Murray), pp. 104–11.

Close, C. (1912) 'The position of geography', *Report of the Eightieth Report of the British Association for the Advancement of Science* (London: John Murray), pp. 436–43.

Cornish, V. (1924) 'The geographical position of the British Empire', *Report of the Ninety-first Meeting of the British Association for the Advancement of Science* (London, Murray), pp. 126–37.

Crawfurd, J. (1862) 'On the connexion between ethnology and physical geography', *Report of the Thirty-first Meeting of the British Association for the Advancement of Science* (London: John Murray), pp. 177–83.

Creak, E. (1904) 'Section E – Geography', *Report of the Seventy-third Meeting of the British Association for the Advancement of Science* (London: John Murray), pp. 701–11.

Daubeny, C. (1837) 'Address by Professor Daubeny', *Report of the Sixth Meeting of the British Association for the Advancement of Science* (London: John Murray, 1836), pp. xxi–xxxvi.

Galton, D. (1895) 'Address', *Report of the Sixty-fifth Meeting of the British Association for the Advancement of Science* (London: John Murray), pp. 3–35.

Galton, F. (1873) 'Transactions of the sections: Geography', *Report of the Forty-second Meeting of the British Association for the Advancement of Science* (London: John Murray), pp. 198–202.

Goldie, G. T. (1906) 'Transactions of the sections: Geography', *Report of the Sixty-seventh Meeting of the British Association for the Advancement of Science* (London: John Murray), pp. 611–17.

Haddon, A. C., and C. R. Browne (1892) 'On the ethnography of the Aran Islands, County Galway', *Proceedings of the Royal Irish Academy* 2, pp. 768–830.

Herbertson, A. J. (1911) 'Geography at the British Association, Portsmouth meeting, 1911', *Geographical Journal* 38, pp. 504–14.

Herschel, J. (1846) 'Address', *Report of the Fifteenth Meeting of the British Association for the Advancement of Science* (London: John Murray), pp. xxvii–xliv.

Hooker, J. (1882) 'Transactions of the sections: Geography', *Report of the Fifty-first Meeting of the British Association for the Advancement of Science* (London: John Murray), pp. 732–8.

Hunt, J. (1865) 'The president's address', *Journal of the Anthropological Society of London* 3, pp. lxxxv–cxiii.

Johnston, D. (1909) 'Section E – Geography', *Report of the Seventy-ninth Meeting of the British Association for the Advancement of Science* (London: John Murray), pp. 517–28.

Keltie, J. S. (1898) 'Section E – Geography', *Report of the Sixty-seventh Meeting of the British Association for the Advancement of Science* (London: John Murray), pp. 699–712.

Lloyd, H. (1859) 'Report to the Council of the British Association as presented to the General Committee at Leeds, September 22nd, 1858', *Report of the Twenty-eighth Meeting of the British Association for the Advancement of Science* (London: John Murray), pp. xxx–xxxi.

Lucas, C. P. (1915) 'Section E – Geography', *Report of the Eighty-fourth Meeting of the British Association for the Advancement of Science* (London: John Murray), pp. 426–39.

Mackinder, H. J. et al. (1898) 'The position of geography in the educational system of the country', *Report of the Sixty-seventh Meeting of the British Association for the Advancement of Science* (London: John Murray), pp. 370–409.

Markham, C. R. (1879) 'Transactions of the sections: Geography', *Report of the Forty-ninth Meeting of the British Association for the Advancement of Science* (London: John Murray), pp. 420–32.

Mill, H. R. (1901) 'On research in geographical science', *Report of the Seventy-first Meeting of the British Association for the Advancement of Science* (London: John Murray), pp. 698–714.

Murchison, R. I. (1839) 'Presidential address', *Report of the Seventh Meeting of the British Association for the Advancement of Science* (London: John Murray), pp. xxxi–xliv.

Murchison, R. I. (1847) 'Presidential address to the British Association for the Advancement of Science', *Report of the Sixteenth Meeting of the British Association for the Advancement of Science* (London: John Murray), pp. xxvii–xliii.

Murchison, R. I. (1859) 'Geography and ethnology', *Report of the Twenty-seventh Meeting of the British Association for the Advancement of Science* (London: John Murray), pp. 143–4.

Murchison, R. I. (1861), 'Geography and ethnology', *Report of the Thirtieth Meeting of the British Association for the Advancement of Science* (London: John Murray), pp. 148–53.
Murchison, R. I. (1864) 'Geography and ethnology', *Report of the Thirty-third Meeting of the British Association for the Advancement of Science* (London: John Murray), pp. 122–3.
Murchison, R. I. (1865) 'Transactions of the sections: Geography and Ethnology', *Report of the Thirty-fourth Meeting of the British Association for the Advancement of Science* (London: John Murray), pp. 130–5.
Murchison, R. I. (1871) 'Geography', *Report of the Fortieth Meeting of the British Association for the Advancement of Science* (London: John Murray), pp. 158–66.
Murchison, R. I., and E. Sabine (1841) 'Address', *Report of the Tenth Meeting of the British Association for the Advancement of Science* (London: John Murray), pp. xxxv–xlviii.
Newbigin, M. (1922) 'Human geography: first principles and some applications', *Report of the Ninety-second Meeting of the British Association for the Advancement of Science* (London: John Murray), pp. 94–105.
Nicholson, C. (1867) 'Geography and ethnology', *Report of the Thirty-sixth Meeting of the British Association for the Advancement of Science* (London: John Murray), pp. 98–9.
Owen, R. (1859) 'Address', *Report of the Twenty-eighth Meeting of the British Association for the Advancement of Science* (London: John Murray), pp. xlix–cx.
Parkinson, T. W. F. (1919) 'Geography in the curriculum of higher education', *Report of the Eighty-seventh Meeting of the British Association for the Advancement of Science* (London: John Murray), pp. 444–8.
Roxby, P. M. (1931) 'The scope and aims of human geography', *Report of the Ninety-eighth Meeting of the British Association for the Advancement of Science* (London: John Murray), pp. 92–204.
Sabine, E. (1837) 'Observations on the direction and intensity of the terrestrial magnetic force in Scotland', *Report of the Sixth Meeting of the British Association for the Advancement of Science* (London: John Murray), pp. 97–119.
Strachey, R. (1876) 'Transactions of the sections: Geography', *Report of the Forty-fifth Annual Meeting of the British Association for the Advancement of Science* (London: John Murray), pp. 180–8.
Turner, A. (1872) 'Anthropology', *Report of the Forty-first Meeting of the British Association for the Advancement of Science* (London: John Murray), pp. 145–7.
Wilson, C. W. (1873) 'Recent surveys in Sinai and Palestine', *Journal of the Royal Geographical Society* 43, pp. 206–40.

British Parliamentary Papers
Memorial of the British Association for the Advancement of Science, in relation to the present state of Survey of Great Britain (1836) XLVII, pp. 89–96.

Newspapers and periodicals
Aris's Birmingham Gazette
Birmingham Post
Bristol Observer
Calgary Daily Herald
Daily Herald
Daily Telegraph
Dublin Evening Post
Dublin University Magazine
Edinburgh Evening Courant
Freeman's Journal and Daily Commercial Advertiser
Free Press News Bulletin
Irish Daily News
Liverpool Daily Post
Leisure Hour
Manchester Courier and Lancashire General Advertiser
Manitoba Free Press
Montreal Gazette
Morning Post
Natal Mercury
Nature
Nottingham Daily Guardian
Punch
Scotsman
Sheffield Daily Telegraph
The Age
The British Critic
The Daily Review
The Irish Times
The Glasgow News
The Nation
The Times
Yorkshire Post

Secondary sources

Agnew, J. and D. Livingstone (eds) (2010), *A Handbook of Geography* (London: Sage).

Akerman, J. R., and R. W. Karrow (2007) (eds) *Maps: Finding our Place in the World* (Chicago: University of Chicago Press).

[Anon.] (1865) 'Anthropology at the British Association', *Anthropological Review* 3, pp. 354–71.

[Anon.] (1871) 'The British Association in Edinburgh', *Scotsman*, 3 August.

[Anon.] (1895) 'Geography at the British Association, Ipswich, 1895', *Geographical Journal* 6, pp. 460–5.

[Anon.] (1897) 'Geography at the British Association', *Geographical Journal* 10, pp. 471–6.
[Anon.] (1898) 'Geography at the British Association, Bristol, 1898', *Geographical Journal* 12, pp. 377–85.
[Anon.] (1909) 'Geography at the British Association', *Scottish Geographical Magazine* 25, pp. 577–85.
[Anon.] (1912) 'Geography at the British Association', *Scottish Geographical Magazine* 28, pp. 571–88.
[Anon.] (1924) 'The British Association meeting in Canada', *Scottish Geographical Magazine* 40, pp. 345–56.
Alberti, S. J. M. M. (2001) 'Amateurs and professionals in one county: biology and natural history in late Victorian Yorkshire', *Journal of the History of Biology* 34, pp. 115–47.
Alberti, S. J. M. M. (2002) 'Placing nature: natural history collections and their owners in nineteenth-century provincial England', *British Journal for the History of Science* 35, pp. 291–311.
Aubin, D. (2003) 'The fading star of the Paris Observatory: astronomers' urban culture of circulation and observation', *Osiris* 18, pp. 79–100.
Barnes, T. (2004) 'Placing ideas: *genus loci*, heterotopia, and geography's quantitative revolution', *Progress in Human Geography* 28, pp. 1–31.
Barton, R. (1987) 'John Tyndall, pantheist: a rereading of the Belfast address', *Osiris* 3, pp. 111–34.
Basalla, G., W. Coleman and R. H. Kargon (eds) (1970) *Victorian Science: A Self-portrait from the Presidential Addresses of the British Association for the Advancement of Science* (Garden City NY: Anchor Books).
Beaver, S. (1982) 'Geography in the British Association for the Advancement of Science', *Geographical Journal* 148, pp. 173–81.
Bell, M., R. A. Butlin and M. Heffernan (eds) (1995) *Geography and Imperialism, 1820–1940* (Manchester: Manchester University Press).
Bensaude-Vincent, B. (2009) 'A historical perspective on science and its "others"', *Isis* 100, pp. 359–68.
Bourguet, M.-N., C. Licoppe and H. Sibum (eds) (2003) *Instruments, Travel and Science: Itineraries of Precision from the Seventeenth to the Twentieth Centuries* (London: Rouledge).
Bridges, R. (2008) 'William Desborough Cooley', in H. Lorimer and C. W. J. Withers (eds), *Geographers Biobibliographical Studies* 27, pp. 43–62.
Brock, W. H. (ed.) (1973) *H. E. Armstrong and the Teaching of Science, 1880–1930* (Cambridge: Cambridge University Press).
Brown, T. N. L. (1971) *The History of the Manchester Geographical Society, 1884–1950* (Manchester: Manchester University Press).
Browne, J. (1983) *The Secular Ark: Studies in the History of Biogeography* (New Haven CT and London: Yale University Press).
Burchfield, J. (1982) 'The British Association and its historians', *Historical Studies in the Physical and Biological Sciences* 13, pp. 165–74.

Burnett, D. G. (2000) *Masters of all they Surveyed: Exploration, Geography and a British El Dorado* (Chicago: University of Chicago Press).

Butlin, R. A. (1995) 'Historical geographies of the British Empire, *c.* 1887–1925', in M. Bell, R. A. Butlin and M. Heffernan (eds) *Geography and Imperialism, 1820–1940* (Manchester: Manchester University Press), pp. 151–88.

Butlin, R. A. (2006) '"Geography: imperial and local": the work of the Royal Geographical Society of Australasia (Queensland Branch), 1885–1945', in R. W. Childs and B. J. Hudson (eds), *Queensland: Geographical Perspectives* (Brisbane: Royal Geographical Society of Queensland), pp. 217–42.

Cahan, D. (2003) 'Looking at nineteenth-century science: an introduction', in D. Cahan (ed.), *From Natural Philosophy to the Sciences: Writing the History of Nineteenth-Century Science* (Chicago: University of Chicago Press, 2003), pp. 3–15.

Cahan, D. (ed.) (2003) *From Natural Philosophy to the Sciences: Writing the History of Nineteenth-Century Science* (Chicago: University of Chicago Press, 2003).

Camerini, J. (1996) 'Wallace in the field', *Osiris* 11, pp. 44–65.

Camerini, J. (1997) 'Remains of the day: early Victorians in the field', in B. Lightman (ed.), *Victorian Science in Context* (Chicago: University of Chicago Press), pp. 354–77.

Cannon. S. F. (1978) *Science in Culture: The Early Victorian Period* (Cambridge MA: Harvard University Press).

Cavell, J. (2008) *Tracing the Connected Narrative: Arctic Exploration in British Print Culture, 1818–1860* (Toronto: University of Toronto Press).

Cawood, J. (1977) 'Terrestrial magnetism and the development of international collaboration in the early nineteenth century', *Annals of Science* 34, pp. 551–87.

Cawood, J. (1979) 'The magnetic crusade: science and politics in early Victorian Britain', *Isis* 70, pp. 493–518.

Childs, R. W., and B. J. Hudson (eds) (2006) *Queensland: Geographical Perspectives* (Brisbane: Royal Geographical Society of Queensland).

Clifford, D., E. Wadge, A. Warwick and M. Willis (eds) (2006) *Repositioning Victorian Sciences: Shifting Centres in Nineteenth-Century Scientific Thinking* (London: Anthem Press).

Cock, R. (2005) 'Scientific servicemen in the Royal Navy and the professionalisation of science, 1816–1855', in D. Knight and M. Eddy (eds), *Science and Beliefs: From Natural Philosophy to Natural Science, 1700–1900* (Aldershot: Ashgate), pp. 95–111.

Collini, S. (2006) *Absent Minds: Intellectuals in Britain* (Oxford: Oxford University Press).

Collins, P. (1981) 'The British Association as public apologist for science, 1919–1946', in R. MacLeod and P. Collins (eds), *The Parliament of Science: The British Association for the Advancement of Science, 1831–1981* (Northwood: Science Reviews), pp. 211–36.

Collins, H. M., and R. Evans (2002) 'The third wave of science studies: studies of expertise and experience', *Social Studies of Science* 32, pp. 235–96.
Collins, H. M., and R. Evans (2007) *Rethinking Expertise* (Chicago: University of Chicago Press).
Collis, C., and K. Dodds (2008) 'Assault on the unknown: the historical and political geographies of the International Geophysical Year, 1957–1958', *Journal of Historical Geography* 34, pp. 555–73.
Cooley, W. D. (1846) 'Synopsis of a proposal respecting a physico-geographical survey of the British Islands, particularly in relation to agriculture', *Report of the Sixteenth Meeting of the British Association for the Advancement of Science* (London: John Murray), pp. 72–3.
Cooter, R., and S. Pumphrey (1994) 'Separate spheres and public spaces: reflections on the history of science popularization and science in popular culture', *History of Science* 32, pp. 237–67.
Cormack, L. (1997) *Charting an Empire: Geography at the English Universities, 1580–1620* (Chicago: University of Chicago Press).
Craggs, R. (2008) 'Situating the imperial archive: the Royal Empire Society Library, 1868–1945', *Journal of Historical Geography* 34, pp. 48–67.
Crawford, E., T. Shinn and S. Sörlin (eds) (1993) *Denationalizing Science: The Contexts of International Scientific Practice* (Dordrecht: Kluwer).
Crawfurd, J. (1863) 'On the connexion between ethnology and physical geography', *Transactions of the Ethnological Society of London* 2, pp. 4–23.
Crowhurst, A. (1997) 'Empire Theatres and the empire: the popular imagination in the Age of Empire', *Environment and Planning D: Society and Space* 15, pp. 155–74.
Dalgleish, W. S. (1894) 'Geography at the British Association, Oxford, August 1894', *Scottish Geographical Magazine* 10, pp. 463–73.
Daston, L. (1991) 'The ideal and the reality of the republic of letters in the Enlightenment', *Science in Context* 4, pp. 367–86.
Daston, L., and P. Galison (2007) *Objectivity* (New York: Zed Books).
Daum, A. W. (2009) 'Varieties of popular science and the transformations of public knowledge: some historical reflections', *Isis* 100, pp. 319–32.
Daunton, M. (ed.) (2005) *The Organisation of Knowledge in Victorian Britain* (London: Oxford University Press in association with the British Academy).
Dawson, S. E. (ed.) (1881) *Handbook for the Dominion of Canada* (Montreal: Dawson Brothers).
Demeritt, D. (2002) 'What is the "social construction of nature"? A typology and sympathetic critique', *Progress in Human Geography* 26, pp. 767–90.
Desmond, A. (2001) 'Redefining the x axis: "professionals", "amateurs" and the making of Victorian biology: a progress report', *Journal of the History of Biology* 34, pp. 3–50.
Desmond, A., and J. Moore (2009) *Darwin's Sacred Cause: Race, Slavery, and the Quest for Human Origins* (London: Allen Lane).
Dettelbach, M. (1996) 'Global physics and aesthetic empire: Humboldt's physical portrait of the tropics', in D. P. Miller and P. H. Reill (eds), *Visions*

of Empire: Voyages, Botany, and Representations of Nature (Cambridge: Cambridge University Press), pp. 258–92.

Dickie, G. (1838) *Flora Abredonensis, comprehending a List of the Flowering Plants and Ferns found in the Neighbourhood of Aberdeen* (Aberdeen, no publisher).

Dierig, S., J. Lachmund and A. J. Mendelsohn (eds) (2003), *Science and the City* (Chicago: University of Chicago Press).

Dierig, S., J. Lachmund and A. J. Mendelsohn (2003) 'Introduction. Toward an urban history of science', *Osiris* 18, pp. 1–19.

Dodds, K.-J. (1994) 'Eugenics, fantasies of empire and inverted Whiggism: an essay on the political geography of Vaughan Cornish', *Political Geography* 13, pp. 85–99.

Dorn, H. (1991) *The Geography of Science* (Baltimore MD: Johns Hopkins University Press).

Drayton, R. (2000) *Nature's Government: Science, Imperial Britain and the 'Improvement' of the World* (New Haven CT: Yale University Press).

Driver, F. (2000) 'Editorial. Fieldwork in geography', *Transactions of the Institute of British Geographers* 25, pp. 267–8.

Driver, F. (2001) *Geography Militant: Cultures of Exploration and Empire* (Oxford: Blackwell).

Dubow, S. (2000) 'A commonwealth of science: the British Association in South Africa, 1905 and 1929', in S. Dubow (ed.), *Science and Society in Southern Africa* (Manchester: Manchester University Press), pp. 66–99.

Dubow, S. (2000) (ed.) *Science and Society in Southern Africa* (Manchester: Manchester University Press).

Dunbar, G. S. (2001) (ed.), *Geography: Discipline, Profession and Subject since 1870: An International Survey* (Dordrecht: Kluwer).

Edmonds, J. A., and R. A. Beardmore (1955) 'John Phillips and the early meetings of the British Association', *The Advancement of Science* 12, pp. 97–104.

Edney, M. (1993) 'Cartography without "progress": reinterpreting the nature and historical development of mapmaking', *Cartographica* 30, pp. 54–68.

Edney, M. (1997) *Mapping an Empire: The Geographical Construction of British India, 1765–1843* (Chicago: University of Chicago Press).

Edney, M. (1998) 'Cartography: disciplinary history', in G. A. Good (ed.), *Sciences of the Earth: An Encyclopedia of Events, People, and Phenomena* (New York and London: Garland), I, pp. 81–5.

Ellen, R. (1988) 'Persistence and change in the relationship between anthropology and human geography,' *Progress in Human Geography* 12, pp. 229–62.

Elliot, G. F. S., M. Laurie and M. J. Barclay (eds) (1901), *Flora, Fauna and Geology of the Clyde Area* (Glasgow: Maclehose).

Endersby, J. (2005) 'Classifying sciences: systematics and status in mid-Victorian natural history', in M. Daunton (ed.), *The Organisation of Knowledge in Victorian Britain* (Oxford: Oxford University Press in association with the British Academy, 2005), pp. 61–86.

Endersby, J. (2008) *Imperial Nature: Joseph Hooker and the Practices of Victorian Science* (Chicago: University of Chicago Press).

Finnegan, D. A. (2004) 'The work of ice: glacial theory and scientific culture in early Victorian Edinburgh', *British Journal for the History of Science* 37, pp. 29–52.

Finnegan, D. A. (2005) 'Natural history societies in late Victorian Scotland and the pursuit of local civic science', *British Journal for the History of Science* 38, pp. 53–72.

Finnegan, D. A. (2007) 'Naturalising the Highlands: geographies of mountain fieldwork in late Victorian Scotland', *Journal of Historical Geography* 33, pp. 791–815.

Finnegan, D. A. (2008)'The spatial turn: geographical approaches in the history of science', *Journal of the History of Biology* 41, pp. 369–88.

Finnegan, D. A. (2009) *Natural History Societies and Civic Culture in Victorian Scotland* (London: Pickering & Chatto).

Fisch, M., and S. Schaffer (eds) (1991) *William Whewell: A Composite Portrait* (Oxford: Clarendon Press).

Fleming, F. (1998) *Barrow's Boys* (London: Granta).

Flint, W., and J. D. F. Gilchrist (eds) (1905) *Science in South Africa: A Handbook and Review* (Cape Town, Pretoria and Bulawayo: T. Maskew Miller).

Forgan, S., and G. Gooday (1996) 'Constructing South Kensington: the buildings and politics of T. H. Huxley's working environments', *British Journal for the History of Science* 29, pp. 435–68.

Freeman, T. W. (1982) 'Charles Bungay Fawcett, 1883–1952', *Geographers Biobibliographical Studies* 6, pp. 39–46.

Freeman, T. W. (1985) 'Charles Frederick Arden-Close, 1865–1952', *Geographers Biobibliographical Studies* 9, pp. 1–14.

Friendly, M., and D. Denis (2000) 'The roots and branches of statistical graphics', *Journal de la Société française de statistique* 141, pp. 51–60.

Friendly, M., and G. Palsky (2007) 'Visualising nature and society', in J. R. Akerman and R. W. Karrow (eds), *Maps: Finding our Place in the World* (Chicago: University of Chicago Press), pp. 207–53.

Fyfe, A., and B. Lightman (2007) 'Science in the marketplace: an introduction', in A. Fyfe and B. Lightman (eds), *Science in the Marketplace: Nineteenth-Century Sites and Experiences* (Chicago: University of Chicago Press), pp. 1–19.

Fyfe, A., and B. Lightman (eds) (2007) *Science in the Marketplace: Nineteenth-Century Sites and Experiences* (Chicago: University of Chicago Press).

Garnett, A. (1983) 'IBG: the formative years: some reflections', *Transactions of the Institute of British Geographers* 8, pp. 27–35.

Gascoigne, J. (1998) *Science in the Service of Empire: Joseph Banks, the British State and the Uses of Science in the Age of Revolution* (Cambridge: Cambridge University Press).

Gieryn, T. (1993) 'Boundary work and the demarcation of science from non-science: strains and interests in the professional ideologies of scientists', *American Sociological Review* 48, pp. 781–95.

Gieryn, T. (1999) *Cultural Boundaries of Science: Credibility on the Line* (Chicago: University of Chicago Press).

Gieryn, T. F. (2005) 'Three truth spots', *Journal of the History of the Behavioural Sciences* 38, pp. 113–32.

Gieryn, T. F. (2006) 'City as truth-spot: laboratories and field-sites in urban studies', *Social Studies of Science* 36, pp. 5–38.

Godlewska, A. M. C. (1999) 'From Enlightenment vision to modern science? Humboldt's visual thinking', in D. Livingstone and C. W. J. Withers (eds), *Geography and Enlightenment* (Chicago: University of Chicago Press), pp. 236–79.

Godlewska, A. M. C. (1999) *Geography Unbound: French Geographic Science from Cassini to Humboldt* (Chicago: University of Chicago Press).

Godlewska, A., and N. Smith (eds) (1994) *Geography and Empire* (London: Blackwell).

Goldman, L. (2002) *Science, Reform and Politics in Victorian Britain: The Social Science Association, 1857–1886* (Cambridge: Cambridge University Press).

Golinski, J. (1998) *Making Natural Knowledge: Constructivism and the History of Science* (Cambridge: Cambridge University Press).

Gooday, G. (2007) 'Illuminating the expert–consumer relationship in domestic electricity', in A. Fyfe and B. Lightman (eds), *Science in the Marketplace: Nineteenth-Century Sites and Experiences* (Chicago: University of Chicago Press), pp. 231–68.

Goren, H. (2002) 'Sacred, but not surveyed: nineteenth-century surveys of Palestine', *Imago Mundi* 54, pp. 87–110.

Gregory, J., and S. Miller (1998) *Science in Public: Communication, Culture and Credibility* (New York: Perseus).

Gruffudd, P. (1994) 'Back to the land: historiography, rurality and the nation in inter-war Wales', *Transactions of the Institute of British Geographers* 19, pp. 61–77.

Guston, T. (2000) *Between Politics and Science: Assuring the Integrity and Productivity of Research* (Cambridge: Cambridge University Press).

Harris, S. (1998) 'Long-distance corporations, big sciences, and the geography of knowledge', *Configurations* 6, pp. 269–305.

Herbertson, A. J. (1905) 'The visit of the British Association to South Africa', *Geographical Journal* 26, pp. 632–41.

Herschel, J. (ed.) (1849) *The Admiralty Manual of Scientific Enquiry* (London: John Murray).

Heyck, T. W. (1982) *The Transformation of Intellectual Life in Victorian England* (London: Croom Helm).

Higgitt, R., and C. W. J. Withers (2008) Science and sociability: women as audience at the British Association for the Advancement of Science', *Isis* 99, pp. 1–27.

Home, R., and S. G. Kohlstedt (eds) (1991) *International Science and National Scientific Identity: Australia between Britain and America* (Dordrecht and London: Kluwer).

Howarth, O. J. R. (1931) *The British Association: A Retrospect* (London: British Association).

Howarth, O. J. R. (1951) 'The centenary of Section E (Geography)', *Advancement of Science* 8, pp. 151–65.

Iliffe, R. (1993) '"Aplatisseur du monde et de cassini" : Maupertuis, precision measurement, and the shape of the earth in the 1730s', *History of Science* 31, pp. 335–75.

James, F. A. L. (2005) 'An "open clash" between science and the Church? Wilberforce, Huxley and Hooker on Darwin at the British Association, Oxford, 1860', in D. M. Knight and M. D. Eddy (eds), *Science and Beliefs: From Natural Philosophy to Natural Science, 1700–1900* (Aldershot: Ashgate), pp. 171–93.

Jeal, T. (2007) *Stanley: The Impossible Life of Africa's Greatest Explorer* (London: Penguin).

Johnston, R. J. (2003), 'The institutionalisation of geography as an academic discipline', in R. Johnston and M. Williams (eds), *A Century of British Geography* (London: Oxford University Press in association with the British Academy), pp. 45–92.

Johnston, R. J., and M. Williams (eds) (2003), *A Century of British Geography* (London: Oxford University Press in association with the British Academy).

Jones, M. (2005) 'Measuring the world: exploration, empire and the reform of the Royal Geographical Society, c. 1874–1893', in M. Daunton (ed.), *The Organisation of Knowledge in Victorian Britain* (London: Oxford University Press in association with the British Academy), pp. 313–35.

Jöns, H. (2008) 'Academic travel from Cambridge University and the formation of centres of knowledge, 1885–1954', *Journal of Historical Geography* 34, pp. 338–62.

Joyce, P. (1991) *Visions of the People: Industrial England and the Question of Class, 1848–1914* (Cambridge: Cambridge University Press).

Kain, R. J. P., and C. Delano-Smith (2003) 'Geography displayed: maps and mapping', in R. Johnston and M. Williams (eds), *A Century of British Geography* (London: Oxford University Press in association with the British Academy), pp. 371–427.

Kaviraj, S., and S. Khilnani (eds) (2001), *Civil Society: History and Possibilities* (Cambridge: Cambridge University Press).

Kearns, G. (2009) *Geopolitics and Empire: The Legacy of Halford Mackinder* (Oxford: Oxford University Press).

Keltie, J. S. (1886) 'Geographical Education: Report to the Council of the Royal Geographical Society', *Supplementary Paper of the Royal Geographical Society* 1, pp. 439–594.

Keltie, J. S. (1886) *Report of the Proceedings of the Royal Geographical Society in Reference to the Improvement of Geographical Education* (London: John Murray).

Kirsch, S. (2002) 'John Wesley Powell and the mapping of the Colorado plateau, 1869–1879: survey science, geographical solutions, and the economy of environmental values', *Annals of the Association of American Geographers* 92, pp. 548–72.

Knibbs, G. H. (ed.) (1914) *Federal Handbook, Prepared in Connection with the Eighty-fourth Meeting of the British Association for the Advancement of Science, held in Australia, August, 1914* (Melbourne: Federal Government).

Knight, D. (2006) *Public Understanding of Science: A History of Communicating Scientific Ideas* (London and New York: Routledge).

Knight, D. M. (1991) 'Tyrannies of distance in British science', in R. W. Home and S. G. Kohlstedt (eds), *International Science and National Scientific Unity: Australia between Britain and America* (Dordrecht: Kluwer), pp. 39–55.

Knight, D. M., and M. D. Eddy (eds) *Science and Beliefs: From Natural Philosophy to Natural Science, 1700–1900* (Aldershot: Ashgate).

Knorr-Cetina, K., and M. Mulkay (eds) (1983) *Science Observed: Perspectives on the Social Study of Science* (Beverly Hills CA and London: University of California Press).

Kohler, R. E. (2002) *Landscapes and Labscapes: Exploring the Lab–Field Border in Biology* (Chicago: University of Chicago Press).

Kraft, A., and S. J. M. M. Alberti (2003) '"Equal though different": laboratories, museums and the institutional development of biology in late Victorian northern England', *Studies in History and Philosophy of Biological and Biomedical Sciences* 34, pp. 203–36.

Kuper, A. (2003) 'Anthropology', in T. Porter and D. Ross (eds), *The Cambridge History of Science*, Vol. 7, *The Modern Social Sciences* (Cambridge: Cambridge University Press), pp. 354–78.

Lane, C. A. (1905) *A Guide to the Transvaal* (Johannesburg: Bartholomew & Lawler).

Latour, B. (1987) *Science in Action: How to Follow Scientists and Engineers through Society* (Cambridge MA: Harvard University Press).

Layton, D. (1981) 'The schooling of science in England, 1854–1939', in R. MacLeod and P. Collins (eds), *The Parliament of Science: The British Association for the Advancement of Science, 1831–1981* (Northwood: Science Reviews), pp. 188–210.

Lefebvre, H. (1991) *The Production of Space*, translated by D. Nicholson-Smith (Oxford: Blackwell).

Lenoir, T. (1997) *Instituting Science: The Cultural Production of Scientific Disciplines* (Stanford CA: Stanford University Press).

Lenoir, T. (ed.) (1998) *Inscribing Science: Scientific Texts and the Materiality of Communication* (Stanford CA: Stanford University Press).

Lightman, B. (ed.) (1997) *Victorian Science in Context* (Chicago: University of Chicago Press).

Lightman, B. (2007) *Victorian Popularizers of Science: Designing Nature for New Audiences* (Chicago: University of Chicago Press).
Livingstone, D. N. (1990) 'Geography', in R. C. Olby, G. N. Cantor, J. R. R. Christie and M. J. S. Hodge (eds), *Companion to the History of Modern Science* (London: Routledge), pp. 743–60.
Livingstone, D. N. (1992) *The Geographical Tradition: Episodes in the History of a Contested Enterprise* (London: Blackwell).
Livingstone, D. N. (1995) 'The spaces of knowledge: contributions towards a historical geography of science', *Environment and Planning D: Society and Space* 13, pp. 5–34.
Livingstone, D. N. (1999) 'Science, region and religion: the reception of Darwin in Princeton, Belfast and Edinburgh', in R. L. Numbers and J. Stenhouse (eds), *Disseminating Darwinism: The Role of Place, Race, Religion, and Gender* (New York: Cambridge University Press), pp. 7–38.
Livingstone, D. N. (2003a) 'British geography, 1500–1900: an imprecise review', in R. Johnston and M. Williams (eds), *A Century of British Geography* (London: Oxford University Press in association with the British Academy), pp. 11–44.
Livingstone, D. N. (2003b) *Putting Science in its Place* (Chicago: University of Chicago Press).
Livingstone, D. N. (2005) 'Text, talk and testimony: geographical reflections on scientific habits: an afterword', *British Journal for the History of Science* 38, pp. 93–100.
Livingstone, D. N. (2007) 'Science, site and speech: scientific knowledge and the spaces of rhetoric', *History of the Human Sciences* 20, pp. 71–98.
Livingstone, D. N. (2008) *Adam's Ancestors: Race, Religion and the Politics of Human Origins* (Baltimore MD: Johns Hopkins University Press).
[Local Publications Committee] (1914) *Handbook and Guide to Western Australia* (Perth WA: W. M. Simpson).
Lochhead, E. (1997) 'The Royal Scottish Geographical Society: the setting and sources of its success', *Scottish Geographical Magazine* 113, pp. 42–50.
Lorimer, H., and N. Spedding (2005) 'Locating field science: a geographical family expedition to Glen Roy, Scotland', *British Journal for the History of Science* 38, pp. 13–34.
Lowe, P. (1976) 'Amateurs and professionals: the institutional emergence of British plant ecology', *Journal of the Society for the Bibliography of Natural History* 7, pp. 517–35.
Lowe, P. (1981) 'The British Association and the provincial public', in R. MacLeod and P. Collins (eds), *The Parliament of Science: The British Association for the Advancement of Science* (Northwood: Science Reviews), pp. 170–87.
Mackenzie, J. M. (1995) 'The provincial geographical societies in Britain, 1884–1914', in M. Bell, R. A. Butlin and M. Heffernan (eds.), *Geography and Imperialism, 1820–1940* (Manchester: Manchester University Press), pp. 93–124.

Mackinder, H. J. (1887) 'On the scope and methods of geography', *Proceedings of the Royal Geographical Society* 9, pp. 141–60.
MacLeod, R. (1981) 'Retrospect: the British Association and its historians', in R. MacLeod and P. Collins (eds), *The Parliament of Science: The British Association for the Advancement of Science, 1831–1981* (Northwood: Science Reviews), pp. 1–16.
MacLeod, R. (1981) 'Introduction. On the advancement of science', in R. MacLeod and P. Collins (eds), *The Parliament of Science: The British Association for the Advancement of Science, 1831–1981* (Northwood: Science Reviews), pp. 17–42.
MacLeod, R. (1996) *Public Science and Public Policy in Victorian England* (Aldershot: Variorum).
MacLeod, R., and Collins, P. (eds) (1981) *The Parliament of Science: The British Association for the Advancement of Science 1831–1981* (Northwood: Science Reviews).
MacLeod, R., J. R. Friday and C. Gregor (1975) *The Corresponding Societies of the British Association for the Advancement of Science, 1883–1929* (London: Mansell).
Maddrell, A. M. C. (1996) 'Empire, emigration and school geography: changing discourses of imperial citizenship, 1880–1925', *Journal of Historical Geography* 22, pp. 373–87.
Maddrell, A. (2009) *Complex Locations: Women's Geographical Work in the UK, 1850–1870* (London: Wiley-Blackwell).
Mandelstam, J. (1994) 'Du Chaillu's stuffed gorillas and the savants from the British Museum', *Notes and Records of the Royal Society of London* 48, pp. 227–45.
Markham, C. R. (1879) 'The mountain passes of the Afghan frontier of British India', *Proceedings of the Royal Geographical Society and Monthly Record of Geography* 1, pp. 38–62.
Matless, D. (1992) 'Regional surveys and local knowledges: the geographical imagination in Britain, 1918–1939', *Transactions of the Institute of British Geographers* 17, pp. 464–80.
Mawer, G. A. (2006) *South by Northwest: The Magnetic Crusade and the Contest for Antarctica* (Edinburgh: Birlinn).
Mayhew, R. (1998) 'Geography in eighteenth-century British education', *Paedagogica Historica* 34, pp. 731–69.
McEwan, C. (1998) Gender, science and physical geography in nineteenth-century Britain', *Area* 30, pp. 215–24.
McFarlane, J. R. (1915) 'Geography at the British Association', *Geographical Journal* 45, pp. 147–51.
McGucken, W. (1979) 'The social relations of science: the British Association for the Advancement of Science,' *Proceedings of the American Philosophical Society* 123, pp. 237–64.
McLaughlin-Jenkins, E. (2001) 'Common knowledge: science and the late Victorian working-class press', *History of Science* 39, pp. 445–65.

Merrill, R. T. (1983) *The Earth's Magnetic Field: Its History, Origin and Planetary Perspective* (London: Academic Press).

Meusburger, P. (2008) 'The nexus of knowledge and space', in P. Meusburger, M. Welker and E. Wunder (eds), *Clashes of Knowledge: Orthodoxies and Heterodoxies in Science and Religion* (Heidelberg: Springer), pp. 35–90.

Meusburger, P., M. Welker and E. Wunder (eds) (2008) *Clashes of Knowledge: Orthodoxies and Heterodoxies in Science and Religion* (Heidelberg: Springer).

Meusburger, P., J. Funke and E. Wunder (2009) 'Introduction. The spatiality of creativity', in P. Meusburger, J. Funke and E. Wunder (eds), *Milieus of Creativity: An Interdisciplinary Approach to Spatiality of Creativity* (Heidelberg: Springer), pp. 1–10.

Meusburger, P (2009) 'Mileus of creativity: the role of places, environments and spatial contexts', in P. Meusburger, J. Funke and E. Wunder (eds), *Milieus of Creativity: An Interdisciplinary Approach to Spatiality of Creativity* (Heidelberg: Springer), pp. 97–154.

Meusburger, P., J. Funke and E. Wunder (eds) (2009) *Milieus of Creativity: An Interdisciplinary Approach to Spatiality of Creativity* (Heidelberg: Springer).

Michie, H., and R. R. Thomas (2003) 'Introduction', in H. Michie and R. R. Thomas (eds), *Nineteenth-Century Geographies: The Transformation of Space: From the Victorian Age to the American Century* (New Brunswick NJ: Rutgers University Press), pp. 1–22.

Michie, H., and R. R. Thomas (2003 eds) *Nineteenth-Century Geographies: The Transformation of Space: From the Victorian Age to the American Century* (New Brunswick NJ: Rutgers University Press).

Mill, H. R. (1887) 'Report to Council', *Scottish Geographical Magazine* 3, pp. 521–30.

Mill, H. R. (1889) 'Report to Council on the British Association meeting at Newcastle, 1889', *Scottish Geographical Magazine* 5, pp. 606–8.

Mill, H. R. (1945) *Life Interests of a Geographer: An Experiment in Autobiography* (East Grinstead: published privately).

Mill, H. R. (1951) 'Geography at the British Association: a retrospect', *Scottish Geographical Magazine* 47, pp. 336–53.

Miller, D. (1996) 'Joseph Banks, empire, and "centres of calculation" in late Hanoverian London', in D. Miller and P. H. Reill (eds), *Visions of Empire: Voyages, Botany, and Representations of Nature* (Cambridge: Cambridge University Press), pp. 21–37.

Miller, D., and P. H. Reill (eds) (1996) *Visions of Empire: Voyagers, Botany, and Representations of Nature* (Cambridge: Cambridge University Press).

Miskell, L. (2003) 'The making of a new "Welsh metropolis": science, leisure and industry in early nineteenth-century Swansea', *History* 88, pp. 32–52.

Morley, D. (1981) 'Yesterday, today and tomorrow, 1945–1981', in R. Macleod and P. Collins (eds), *The Parliament of Science: The British Association for the Advancement of Science, 1831–1981* (Northwood: Science Reviews), pp. 237–60.

Morrell, J. B. (2005) *John Phillips and the Business of Victorian Science* (Aldershot: Ashgate).

Morrell, J. B., and A. Thackray (1981) *Gentlemen of Science: Early Years of the British Association for the Advancement of Science* (Oxford: Clarendon).

Morrell, J. B., and A. Thackray (1984) *Gentlemen of Science: Early Correspondence of the British Association for the Advancement of Science* (London: Royal Historical Society).

Morton, G., B. de Vries and R. J. Morris (eds) (2006) *Civil Society, Associations and Urban Places: Class, Nation and Culture in Nineteenth-Century Europe* (Aldershot: Ashgate).

Naylor, S. (2002) 'The field, the museum and the lecture hall: the spaces of natural history in Victorian Cornwall', *Transactions of the Institute of British Geographers* 27, pp. 494–513.

Naylor, S. (2005) 'Introduction. Historical geographies of science: places, contexts, cartographies', *British Journal for the History of Science* 38, pp. 1–12.

Newbigin, M. (1929) 'The South African meeting of the British Association', *Scottish Geographical Magazine* 45, pp. 337–49.

Numbers, R. L., and J. Stenhouse (eds) (1999) *Disseminating Darwinism: The Role of Place, Race, Religion, and Gender* (New York: Cambridge University Press).

Nyhart, L. K., and T. H. Broman (eds) (2002) *Science and Civil Society* (Chicago: University of Chicago Press).

O'Connor, R. (2009) 'Reflections on popular science in Britain: genres, categories, and historians', *Isis* 100, pp. 333–45.

Olby, R. C., G. N. Cantor, J. R. R. Christie and M. J. S. Hodge (eds) (1990) *Companion to the History of Modern Science* (London: Routledge).

Ophir, A., and S. Shapin (1991) 'The place of knowledge: a methodological survey', *Science in Context* 4, pp. 3–21.

Opitz, D. (2004) '"Behind folding shutters in Whittinghame House": Alice Blanche Balfour, 1864–1936, and amateur natural history', *Archives of Natural History* 31, pp. 330–48.

Orange, A. D. (1971) 'The British Association for the Advancement of Science: the provincial background', *Science Studies* 1, pp. 315–29.

Overy, R. (2009) *The Morbid Age: Britain between the Wars* (London: Allen Lane).

Pancaldi, G. (1981) 'Scientific internationalism and the British Association', in R. MacLeod and P. Collins (eds), *The Parliament of Science: The British Association for the Advancement of Science, 1831–1981* (Northwood: Science Reviews), pp. 145–69.

Pandora, K. (2009) 'Popular science in national and transnational perspective: suggestions from the American context', *Isis* 100, pp. 346–58.

Paton A. W., and A. H. Millar (1912) *Handbook and Guide to Dundee and District* (Dundee: Caird).

Pickering, A. (ed.) (1982) *Science as Practice and Culture* (Chicago: University of Chicago Press).

Pickstone, J. V. (2000) *Ways of Knowing: A New History of Science, Technology and Medicine* (Manchester: Manchester University Press).

Ploszajska, T. (1998) 'Down to earth? Geography fieldwork in English schools, 1870–1914', *Environment and Planning D: Society and Space* 16, pp. 757–74.

Ploszajska, T. (1999) *Geographical Education, Empire and Citizenship: Geographical Teaching and Learning in English Schools, 1870–1944* (Cambridge: HGRG Publications No. 35).

Poovey, M. (1998) *A History of the Modern Fact: Problems of Knowledge in the Sciences of Wealth and Society* (Chicago: University of Chicago Press).

Porter, A. (ed.) (1999) *The Oxford History of the British Empire: The Nineteenth Century* (Oxford: Oxford University Press).

Porter, R. (ed.) (2003) *The Cambridge History of Science*, Vol. 4, *Eighteenth-Century Science* (Cambridge: Cambridge University Press).

Porter, T. (1995) *Trust in Numbers: The Pursuit of Objectivity in Science and Public Life* (Princeton NJ: Princeton University Press)

Porter, T. (2003) 'The social sciences', in D. Cahan (ed.), *From Natural Philosophy to the Sciences: Writing the History of Nineteenth-Century Science* (Chicago: University of Chicago Press), pp. 254–90.

Porter, T., and D. Ross (eds) (2003) *The Cambridge History of Science*, Vol. 7, *The Modern Social Sciences* (Cambridge: Cambridge University Press).

Powell, R. (2007) 'Geographies of science: histories, localities, practices, futures', *Progress in Human Geography* 31, pp. 309–30.

[Publication Committee] (1897) *Handbook of Canada* (Toronto: Publication Committee of the Local Executive).

Pumfrey, S. (2002) *Latitude and the Magnetic Earth* (Cambridge: Icon Books).

Pyenson, L. (2002) 'An end to national science: the meaning and the extension of local knowledge', *History of Science* 40, pp. 251–90.

Pym, H. N. (ed.) (1882) *Memories of Old Friends, being Extracts from the Journals and Letters of Caroline Fox* (London: Smith Elder).

Raj, K. (2003) 'When human travellers become instruments: the Indo-British exploration of Central Asia in the nineteenth century', in M.-N. Bourguet, C. Licoppe and H. Sibum (eds), *Instruments, Travel and Science: Itineraries of Precision from the Seventeenth to the Twentieth Centuries* (London: Routledge), pp. 156–88.

Raj, K. (2007) *Relocating Modern Science: Circulation and the Construction of Knowledge in South Asia and Europe, 1650–1900* (Basingstoke: Palgrave).

Raleigh, Lady [Clara Strutt] (1884) *The British Association's Visit to Montreal, 1884: Letters by Clara Lady Raleigh* (London: printed privately).

Ravenstein, E. G. (1886) 'On bathy-hypsographical maps, with special reference to a combination of the Ordnance and Admiralty charts', *Proceedings of the Royal Geographical Society and Monthly Record* 8, pp. 21–7.

Reidy, M. (2008) *Tides of History: Ocean Science and Her Majesty's Navy* (Chicago: University of Chicago Press).

Richards, T. (2003) *The Imperial Archive: Knowledge and the Fantasy of Empire* (London: Verso).

Richardson, R. L. (1884) *Report of the British Association to the Canadian North West: Description of the Trips to the Rocky Mountains, Addresses Presented, Report on Speeches Delivered, Doings in Winnipeg* (Winnipeg: no publisher).

Ritvo, H. (1997) 'Zoological nomenclature and the empire of Victorian science', in B. Lightman (ed.), *Victorian Science in Context* (Chicago: University of Chicago Press), pp. 334–53.

Robic, M.-C. (2003) 'Geography', in T. Porter and D. Ross (eds), *The Cambridge History of Science*, Vol. 7, *The Modern Social Sciences* (Cambridge: Cambridge University Press), pp. 379–90.

Rudmose Brown, R. (1914) 'The British Association in Australia', *Scottish Geographical Magazine* 30, pp. 631–5.

Rudmose Brown, R. N. (1923) *A Naturalist at the Poles* (London: Seeley Service).

Rudwick, M. J. S. (2005) *Bursting the Limits of Time* (Chicago: University of Chicago Press).

Rudwick, M. J. S. (2008) *Worlds before Adam* (Chicago: University of Chicago Press).

Rupke, N. A. (ed.) (2002) *Göttingen and the Development of the Natural Sciences* (Göttingen: Wallstein).

Rusnock, A. (2002) *Vital Accounts: Quantifying Health and Population in Eighteenth-Century England and France* (Cambridge: Cambridge University Press).

Russell, C. A. (1983) *Science and Social Change, 1700–1900* (London: Macmillan).

Ryan, J. R. (1997) *Picturing Empire: Photography and the Visualisation of the British Empire* (London: Reaktion).

Scargill, P. I. (1976) 'The RGS and the foundations of geography at Oxford', *Geographical Journal* 142, pp. 438–61.

Schaffer, S. (1991) 'The history and geography of the intellectual world: Whewell's politics of language', in M. Fisch and S. Schaffer (eds), *William Whewell: A Composite Portrait* (Oxford: Clarendon Press), pp. 201–31.

Schaffer, S. (1998) 'Physics laboratories and the Victorian country house', in C. Smith and J. Agar (eds), *Making Space for Science: Territorial Themes in the Making of Knowledge* (Basingstoke: Macmillan), pp. 149–80.

Schuster, J., and R. Yeo (eds) (1986) *The Politics and Rhetoric of Scientific Method: Historical Studies* (Dordrecht and Boston MA: Kluwer Academic).

Schwartz, J. (1996) '*The Geography Lesson*: photographs and the construction of imaginative geographies', *Journal of Historical Geography* 22, pp. 16–45.

Schulten, S. (2001) *The Geographical Imagination in America, 1880–1950* (Chicago: University of Chicago Press).

Secord, A. (1994) 'Scientists in the pub: artisan botanists in early nineteenth-century Lancashire', *History of Science* 32, pp. 269–315.

Secord, J. A. (1982) 'King of Siluria: Roderick Murchison and the imperial theme in nineteenth-century British geology', *Victorian Studies* 25, pp. 413–42.
Secord, J. A. (2000) *Victorian Sensation: The Extraordinary Publication, Reception, and Secret Authorship of* Vestiges of the Natural History of Creation (Chicago: University of Chicago Press).
Secord, J. A. (2004) 'Knowledge in transit', *Isis* 95, pp. 654–72.
Secord, J. A. (2007) 'How scientific conversation became shop talk', *Transactions of the Royal Historical Society*, sixth series, XVII, pp. 129–56.
Secord, J. A. (2007) 'How scientific conversation became shop talk', in A. Fyfe and B. Lightman (eds), *Science in the Marketplace* (Chicago: University of Chicago Press), pp. 23–59.
Seymour, W. A. (ed.) (1978) *A History of the Ordnance Survey* (Folkestone: Dawson).
Shapin, S. (1990) 'Science and the public', in R. Olby, G. Cantor, M. J. S. Hodge and J. R. R. Christie (eds), *The Companion to Modern Science* (London: Routledge), pp. 990–1001.
Shapin, S. (1998) 'Placing the view from nowhere: historical and sociological problems in the location of science', *Transactions of the Institute of British Geographers* 23, pp. 5–12.
Shteir, A. (2007) 'Sensitive, bashful, and chaste? Articulating the *Mimosa* in science', in A. Fyfe and B. Lightman (eds), *Science in the Marketplace: Nineteenth-Century Sites and Audiences* (Chicago: University of Chicago Press), pp. 169–96.
Simoes, A., A. Carneiro and M. P. Diogo (eds) (2003) *Travels of Learning: A Geography of Science in Europe* (Dordrecht: Kluwer).
Simpson, J. Y. (1905) 'The South African meeting of the British Association', *Scottish Geographical Magazine* 20, pp. 637–52.
Smith, C., and J. Agar (eds) (1998) *Making Space for Science: Territorial Themes in the Making of Knowledge* (Basingstoke: Macmillan).
Somerville, M. (1848) *Physical Geography* (London: John Murray).
Sorrenson, R. (1996) 'The ship as a scientific instrument in the eighteenth century', *Osiris* 11, pp. 221–36.
Spary, E. C. (2000) *Utopia's Gardens: French Natural History from Old Regime to Revolution* (Chicago: University of Chicago Press).
Speak, P. (1997) 'Williams Spiers Bruce, 1867–1921', *Geographers Biobibliographical Studies* 17, pp. 17–25.
Stafford, R. (1989) *Scientist of Empire: Sir Roderick Murchison, Scientific Exploration and Victorian Imperialism* (Cambridge: Cambridge University Press).
Stafford, R. (1999) 'Scientific exploration and empire', in A. Porter (ed.), *The Oxford History of the British Empire: The Nineteenth Century* (Oxford: Oxford University Press), pp. 294–319.
Steel, R. W. (1983) *The Institute of British Geographers: The First Fifty Years* (London: Institute of British Geographers).

Stevens, A. (1939) 'Address to Section E – Geography', *Report of the Dundee Meeting* (London: British Association), pp. 204–25.
Stocking, G. (ed.) (1984) *Functionalism Historicised: Essays on British Social Anthropology* (Madison WI: University of Wisconsin Press).
Stocking, G. (1987) *Victorian Anthropology* (New York: Free Press).
Stoddart, D. R. (1975) 'The RGS and the foundations of geography at Cambridge', *Geographical Journal* 141, pp. 216–39.
Stoddart, D. (1986) *On Geography* (Oxford: Blackwell).
Stoler, A. (2002) 'Colonial archives and the arts of governance', *Archival Science* 2, pp. 87–109.
Taylor, G. (1914) 'The physical and general geography of Australia', in G. H. Knibbs (ed.), *Federal Handbook, Prepared in Connection with the Eighty-fourth Meeting of the British Association for the Advancement of Science, held in Australia, August, 1914* (Melbourne: Federal Government), pp. 86–121.
Terrall, M. (2003) *The Man who Flattened the Earth: Maupertuis and the Sciences in the Enlightenment* (Chicago: University of Chicago Press).
Topham, J. R. (2009) 'Introduction', *Isis* 100, pp. 310–18.
Turner, F. M. (1980) 'Public science in Britain', *Isis* 71, pp. 589–608.
Turner, F. M. (1993) *Contesting Cultural Authority: Essays in Victorian Intellectual Life* (Cambridge: Cambridge University Press).
Urry, J. (1984) 'Englishmen, Celts and Iberians: the Ethnographic Survey of the United Kingdom, 1892–1899', in G. Stocking (ed.), *Functionalism Historicised: Essays on British Social Anthropology* (Madison WI: University of Wisconsin Press, 1984), pp. 83–105.
Walford, R. (2001) *Geography in British Schools, 1850–2000* (London: Woburn Press).
Walters, A. (1997) 'Conversation pieces: science and polite society in eighteenth-century England,' *History of Science* 35, pp. 121–54.
Warwick, A. (2006) 'Margins and centres', in D. Clifford, E. Wadge, A. Warwick and M. Willis (eds), *Repositioning Victorian Sciences: Shifting Centres in Nineteenth-Century Scientific Thinking* (London: Anthem Press), pp. 1–13.
Whewell, W. (1837) *History of the Inductive Sciences: From the Earliest to the Present Times*, 3 vols (London: J. W. Parker).
Wilson, C. W. (1899) 'Address delivered at the annual meeting of the Fund', *Palestine Exploration Fund Quarterly Statement*, pp. 304–16.
Winichakul, T. (1994) *Siam Mapped: A History of the Geo-body of a Nation* (Honolulu HI: University of Hawaii Press).
Winlow, H. (2001) 'Anthropometric cartography: constructing Scottish racial identity in the early twentieth century', *Journal of Historical Geography* 27, pp. 507–28.
Withers, C. W. J. (2000) 'Authorizing landscape: "authority", naming and the Ordnance Survey's mapping of the Scottish Highlands in the nineteenth century', *Journal of Historical Geography* 26, pp. 532–54.

Withers, C. W. J. (2001) 'A partial biography: the formalization and institutionalization of geography in Great Britain since 1887', in G. S. Dunbar (ed.), *Geography: Discipline, Profession and Subject since 1870: An International Survey* (Dordrecht: Kluwer), pp. 79–119.
Withers, C. W. J. (2001) *Geography, Science and National Identity: Scotland since 1520* (Cambridge: Cambridge University Press).
Withers, C. W. J. (2002) 'The geography of scientific knowledge', in N. A. Rupke (ed.), *Göttingen and the Development of the Natural Sciences* (Göttingen: Wallstein), pp. 9–18.
Withers, C. W. J. (2007) *Placing the Enlightenment: Thinking Geographically about the Age of Reason* (Chicago: University of Chicago Press).
Withers, C. W. J. (2010) 'Geography's intellectual traditions', in J. Agnew and D. Livingstone (eds), *A Handbook of Geography* (London: Sage), forthcoming.
Withers, C. W. J., and R. J. Mayhew (2002) 'Rethinking "disciplinary" history: geography in British universities, c. 1580–1887', *Transactions of the Institute of British Geographers* 27, pp. 11–29.
Withers, C. W. J., and D. A. Finnegan (2003) 'Natural history societies, fieldwork and local knowledge in nineteenth-century Scotland', *Cultural Geographies* 10, pp. 334–53.
Withers, C. W. J., D. A. Finnegan and R. Higgitt (2006) 'Geography's other histories? Geography and science in the British Association for the Advancement of Science, 1831–c. 1933', *Transactions of the Institute of British Geographers* 31, pp. 433–51.
Withers, C. W. J., R. Higgitt and D. A. Finnegan (2008) 'Historical geographies of provincial science: themes in the setting and reception of the British Association for the Advancement of Science in Britain and Ireland, 1831–c. 1939', *British Journal for the History of Science* 41, pp. 385–415.
Worboys, M. (1981) 'The British Association and empire: science and social imperialism', in R. MacLeod and P. Collins (eds.), *The Parliament of Science: The British Association for the Advancement of Science, 1831–1981* (Northwood: Science Reviews), pp. 170–87.
Worster, D. (2001) *A River Running West: The Life of John Wesley Powell* (Oxford: Oxford University Press).
Yearley, S. (2005) *Making Sense of Science* (London: Sage).
Yeo, R. (1986) 'Scientific method and the rhetoric of science in Britain, 1830–1917', in J. Schuster and R. Yeo (eds), *The Politics and Rhetoric of Scientific Method: Historical Studies* (Dordrecht and Boston MA: Kluwer), pp. 259–97.
Yeo, R. (1993) *Defining Science: William Whewell, Natural Knowledge and Public Debate in Victorian Britain* (Cambridge: Cambridge University Press).
Yeo, R. (2001) *Science in the Public Sphere: Natural Knowledge in British Culture, 1800–1860* (Aldershot and Burlington VT: Ashgate).
Yost, R. M. (1999) 'Pondering the imponderable: John Robison and magnetic theory in Britain, c. 1775–1805', *Annals of Science* 56, pp. 143–74.

Index

Note: page numbers in *italics* refer to illustrations; 'n' after a page reference indicates the number of a note on that page.

Aberdeen
  1859 BAAS meeting in 25, 47, 158
  1885 BAAS meeting in 25, 48, 82, 184
  1934 BAAS meeting in 84
Admiralty Manual of Scientific Enquiry (Herschel) 144–5, 170
Africa 8, 51, 66, 70, 71, 77, 81, 83, 84, 92, 174, 190
agriculture 41, 118, 217
  see also section M (agriculture)
Antartic 71, 74, 143, *144*, 148–9, 152, 159, 169
anthropology
  relations of with ethnology 165–82
  relations of with geography 80, 165–82, 227n.20
  see also section H (anthropology)
anthropometry 170, 177
Arctic 152, 155, 159, 174
audiences
  at BAAS meetings xv, xvii, 1–2, 13–14, 28–9, 44–54, 55–7, 64n.55, 88–95, 211–12, 224–5, 235–8
  proportion of in BAAS meetings 47–9, 55–7, 124, 171, 175, 224
Australia
  1914 BAAS meeting in 24, 25, 41, 44, 49, 71, 79–80, 83, 103, 112–14, 118–21, 124, 129

Barker, W. H. 195n.34, 214–15
Bath
  1864 BAAS meeting in 25, 47, 81, 85, 87, 92, *93*, 190
  1888 BAAS meeting in 25, 48, 82
Beckit, H. O. 74, 75, 187
Belfast
  1852 BAAS meeting in 25, 47, 171
  1874 BAAS meeting in 12, 25, 47, 81, 152, 217
  1902 BAAS meeting in 25, 48, 82, 119
Birmingham
  1839 BAAS meeting in 13, 24, 25, 47, 168, 190
  1849 BAAS meeting in 25, 47, 173
  1865 BAAS meeting in 4, *5*, 25, 45, 47, 81, 82, *172*, 175, 176, 190
  1886 BAAS meeting in 25, 39, 48
  1913 BAAS meeting in 25, 49, 83, 216
Blackpool
  1936 BAAS meeting in 84
botanical geography 80, 137–8
botany 15, 41, 56, 67, 76, 113, 118, 137, 140, 149, 205, 217, 245
  geography and 15, 76, 113, 137, 140
  see also section K (botany)
Bournemouth
  1919 BAAS meeting in 25, 49, 83, 119, 213
Bradford
  1873 BAAS meeting in 2, 25, 48, 81

Index

1900 BAAS meeting in 25, 48, 82, 119
Brewster, Sir David 56, 158
Brighton
  1872 BAAS meeting in 25, 48, 51, 81, 85, 92, 153–5, 185
British Association for the Advancement of Science *see* sections A-M
Bristol
  1836 BAAS meeting in 24, 25, 36, 47
  1875 BAAS meeting in 25, 47, 56, 81, 86
  1898 BAAS meeting in 25, 40, 48, 82, 190
  1930 BAAS meeting in 25, 49, 84, 245n.7
British Science Guild 212, 228n.42

Cambridge
  1833 BAAS meeting in 24, 25, 26, 47
  1845 BAAS meeting in 25, 46, 47, 146
  1862 BAAS meeting in 25, 47
  1904 BAAS meeting in 25, 48, 93
  1938 BAAS meeting in 25, 84
  University of 8, 35, 183, 184, 186, 238
Canada
  geography and empire in 79–80, 106
Cardiff
  1891 BAAS meeting in 25, 48, 82
  1920 BAAS meeting in 25, 49, 83
Challenger, HMS 80, 81, 82, 147, 149
Cheltenham
  1856 BAAS meeting in 25, 47, 229n.57
chemistry 69, 202, 203
  see also section B (chemistry)
Chisholm, George Goudie 73, 74, 83, 207–8
citizenship
  geography and 76, 199–202, 212–16, 240, 242–3
civic science, idea of xv, 3–9, 24–61, 233–7
climatology 72, 73, 78, 113, 119, 140, 227n.20

Close, Sir Charles 74, 75, 83, 87, 204–9, *205*, 215, 221, 224, 241
colonialism
  geography and 9–10, 103–29, 213–16
  science and 9–10, 103–29, 213–16
Committee on the Human Geography of Tropical Africa 75, 119, 181–2
Conference of Corresponding Societies 26–7, 27, 40, 101n.37, 218, 230n.62
conversazione 11, 14, 27, 39, 43, *59*, 88
Cork
  1843 BAAS meeting in 25, 47
Cornish, Vaughan 83, 117, 215–16
craniometry 53, 177
Crawfurd, John 68, 81, 92, 171, 173–4, 192
Creak, Capt. E. 82, 149

Darwin, Charles 92, 152, 168, 169, 185, 192
Davis, W. M. 122, 208–9
Debenham, Frank 76, 84, 187, 188, 223, 225
display, practices of in science 6, 9–10, 12, 38–42
Dover
  1899 BAAS meeting in 25, 82, 119
Dublin
  1835 BAAS meeting in 24, 25, 47, 51, 142, 145, 160n.14
  1857 BAAS meeting in 25, 42–3, 47, 53–4, 55, 91–2, 147, 150, 177
  1878 BAAS meeting in 14–15, 25, 35, 43, 48, 81
  1908 BAAS meeting in 53, 92
Dubow, Saul 11, 105, 113–14, 127
Du Chaillu, Paul 92, 189, *190*
Dundee
  1867 BAAS meeting in 2, 25, 35, 36, 47, 81
  1912 BAAS meeting in 25, 44, 49, 83, 87–8, 209–12, 217
  1939 BAAS meeting in 24, 25, 49, 84
  Chamber of Commerce 148

economics 15, 69, 80
  see also section F (economics)
Edinburgh
  1834 BAAS meeting in 24, 25, 36, 47, 142, 150–1, 155
  1850 BAAS meeting in 25, 47
  1871 BAAS meeting in 1–2, 25, 48, 60, 81
  1892 BAAS meeting in 25, 36, 37, 43, 48, 59, 60, 82, 237
  1921 BAAS meeting in 25, 36, 49, 83
  University of 1, 207
education 41, 71, 73, 74
  geography and 41, 71, 73, 74, 78, 82, 84, 183–7, 198–226
  see also section L (education)
empire
  geography and 8, 10–11, 19n.15, 41, 72–4, 79, 82–3, 103–29, 199–200, 213–17, 240
  science and 9–11, 79–80, 82–3, 103–29, 199–200
ethnography 167, 177, 227n.20
  see also ethnology
ethnology 43, 67, 77–8, 84, 165–93, 179, 180
  see also anthropology; section H (anthropology)
excursions 13, 28, 42–4, 45, 58, 122, 124–6, 225, 234
Exeter
  1869 BAAS meeting in 25, 48, 181
exploration 7–8, 66, 72, 73, 77, 78, 79–80, 81–9, 127, 136, 142–50, 155–8, 165–93, 206, 221, 222, 240

Faraday, Michael 15, 56
Fawcett, C. B. 75, 84, 119, 187, 213, 219
fieldwork 28, 39–40, 43–4, 71–5, 145, 156, 177–82, 234
  see also exploration
Fleure, H. J. 84, 213–15, 230n.62
Flora Antarctica (Hooker, J.) 140, *144*
Franklin, Sir John 90, 147, 148, 156

Galton, Francis 81, 85, 86, 90, 92, 155, 185

Garnett, Alice 66–7
geodesy 136, 149–50, 155–6, 165
  see also mapping, geodetic; terrestrial magnetism
Geographical Association 71, 183, 215, 218, 230n.62
Geographical Journal 77, 91, 205, 221
Geographical Magazine 91
geography
  as a science of distribution 70, 80, 81, 135–59
  botany and 15, 76, 113, 137, 140
  citizenship and 76, 199–202, 212–16, 240, 242–3
  colonialism and 9–10, 103–29, 213–16
  considered the 'Ladies' section' 15, 93, 175, 177
  definitions of 8–9, 123–4, 139, 158–9, 206–9, 211, 223, 233, 241–5
  education and 78, 82, 84, 183–7, 198–226
  empire and 8, 10–11, 19n.15, 41, 72–4, 79, 82–3, 103–29, 199–200, 213–17, 240
  human 83, 84, 170, 183–93, 221–3, 240
  in British university education 8–9, 74, 82, 97, 213, 223–4
  mapping and 8–9, 40, 71, 72–4, 78, 81–4, 136, 140–4, 150–7, 172, 181–2
  of science xv, 1–16, 24–61, 68–77, 232–45
  scientific method in 81–6
  teaching of in Dominion universities 74, 119, 133n.52
  thematic specialisation in 41, 78–80, 84, 97, 182, 186, 212, 221–2, 227n.20
  see also section E (geography); 'scientific geography', idea of
geology 30, 38, 40–1, 56, 80, 118, 132n.34, 149, 205, 217, 239
  see also section C (geology)
geomorphology 79, 80, 212
Gill, Sir David 107, 113

Glasgow
  1840 BAAS meeting in 25, 47, 146, 152
  1855 BAAS meeting in 25, 47
  1876 BAAS meeting in 25, 48, 50, 81
  1901 BAAS meeting in 25, 43–4, 48, 50, 60, 82, 119, 217, 234
  1928 BAAS meeting in 25, 49, 84
gorillas 92, *190*
Greenough, George 29–30, 98n.6
Gregory, J. W. 83, 124, 207

handbooks (BAAS) 12–14, 38–9, 40, 58, 91, 112–14, 216–23, 234
Herbertson, A. J. 74, 83, 116, 121, 185, 201–3, 208–9
Herdman, W. A. 110, 118, 132n.34, 203–4
Herschel, John 15, 56, 143, 144–5, 149, 170
Himalayas 71, 77, 82
Hinks, A. R. 83, 221, 223, 230n.73
Hooker, Sir Joseph 80, 81, 140, *144*, 148, 158
Howard, Lady Caroline 42–3, 55–6, 91–2
Howarth, O. J. R. 10, 66, 69–71, 118, 171, 201–3, 209–11, 214
Hull
  1853 BAAS meeting in 25, 32–3, 36, 47, 62n.21
  1922 BAAS meeting in 25, 41, 49, 83, 218–19
Humboldt, Alexander von 137, 158
'Humboldtian science', idea of 137, 145, 156–7
Hunt, James 175–7
Huxley, Thomas Henry 12, 15, 152, 192
hydrography 70, 72, 82, 119, 145, *154*, 160n.14, 227n.20

imperialism 8, 41, 103–29
  see also colonialism; empire
Institute of British Geographers 7, 199, 223
International Geographical Congress (1895) 153, 185

Ipswich
  1851 BAAS meeting in 25, 44, 47
  1895 BAAS meeting in 25, 48, 82, 153, 185
Ireland xv, 2, 24, 28, 111, 151, 233

Keltie, John Scott 73, 82, 96, 116, 120, 183–7, 201–3, 207–9, 224, 240

lecturing 3, 11–12, 29–30, 36, 111
  see also section E (geography) presidential addresses; speech
Leeds
  1858 BAAS meeting in 25, 31–2, 36, 47, 81
  1890 BAAS meeting in 25, 40, 48, 82
  1927 BAAS meeting in 25, 49, 60, 84, 219, *220*, 245n.7
Leicester
  1907 BAAS meeting in 25, 49, 83
  1933 BAAS meeting in 25, 49, 84, *224*
Liverpool
  1837 BAAS meeting in 24, 25, 47
  1854 BAAS meeting in 25, 47
  1870 BAAS meeting in 25, 48, 81
  1896 BAAS meeting in 25, 36, 48, 82, 85
  1923 BAAS meeting in 25, 41, 44, 49, 83, 215–16, 234, 245n.7
Livingstone, David 15, 51, *53*, 55, 91–2, 93, 156, 243
localist emphases in BAAS science 4, 38–42, 200–12, 216–23
Lloyd, Humphrey 143, 147–8
London
  1931 BAAS meeting in 24, 25, 44, 66, 84
Lucas, Sir Charles P. 71, 74, 75, 83, 113, 119, 120–1, 181–2, 192, 195n.34
Lyell, Sir Charles 34, 158

McFarlane, J. R. 75, 83, 187, 195n.34
Mackinder, Halford J. 70, 82, 84, 96, 183, 201–3, 214, 225
MacLeod, Roy 10, 105, 198, 200, 212, 232

'Magnetic Crusade' 136, 140–50
  see also terrestrial magnetism
Manchester
  1842 BAAS meeting in 25, 47
  1861 BAAS meeting in 25, 42, 47, 81, 92, 190, 234
  1887 BAAS meeting in 25, 48, 82, 91, *138*
  1915 BAAS meeting in 25, 49, 83
  Geographical Society 101n.37, 213
mapping
  as a science of empire 8–9, 150–7, 181
  geodetic 72, 77, 82, 136, 140–50
  geography and 8–9, 40, 71–4, 78, 81–4, 136, 140–4, 150–5, 172, 181–2
  hydrographic 72, *154*
  topographic 71, 81, 83, 150–7
  warfare and 155–6
Markham, Sir Clements R. 73, 81, 76–7, 91, 147–8, 150, 152–3, 155–6, 185, 207, 209, 240
mathematics 70, 76, 136–7, 245
  see also section A (mathematics and physics)
meteorology 113, 149, 155, 227n.20
Mill, Hugh Robert 66–7, 73, 82, 87, 88–9, 91, 95, 117, 139, 156, 185, 208–9, 223, 225, 240
Montreal
  1884 BAAS meeting in 24, 25, 48, 79, 82, 103, 106, 107–8, 119, 178
Morrell, Jack 9–10, 24–5, 26, 55, 136
Murchison, Sir Roderick Impey
  on geography and ethnology 85, 87, 171, *172*, 173–7, 189
  presidential addresses of 81, 87, 137, 146–7
  proponent of geography 29, 83, 85, 136–9, 150–1, 155, 157–8, 165, 173, 189, 240
  proponent of geology 138–9
  proponent of terrestrial magnetism 143, 146, 148–9
Murray, Sir John 71, 73, 74, 82
museums 6, 34, 38, 58
Myres, J. L. 74, 84, 218–19, 230n.62

Nansen, Fridtjof 88, 243
Natal 118–19, 133n.48
National Committee for Geography 213, 229n.45
Newbigin, Marion 83, *188*, 192–3, 221
Newcastle
  1838 BAAS meeting in 24, 25, 26, 29, 40, 47, 143
  1863 BAAS meeting in 25, 33–4, 40, 47, 60, 81
  1889 BAAS meeting in 25, 40–1, 48, 82
  1916 BAAS meeting in 25, 49, 83
New Guinea 71, 72, 119
newspapers 1–2, 13–14, 49–50, 215–16
Norwich
  1868 BAAS meeting in 25, 34, 44, 47, 81, 84
  1935 BAAS meeting in 225
Nottingham
  1866 BAAS meeting in 25, 47, 140, 190
  1893 BAAS meeting in 25, 48, 82
  1937 BAAS meeting in 25, 84
Nunn, T. Percy 71, 74, 75, 214

oceanography 78, 132n.34, 227n.20
Ogilvie, A. G. 84, 187, 195n.34
Ordnance Survey 71, 91, 140, 151–2, 204
Oxford
  1832 BAAS meeting in 24, 25, 47
  1847 BAAS meeting in 25, 47
  1860 BAAS meeting in 12, 25, 47, 81, 87, 90, 148, 152, 158
  1894 BAAS meeting in 25, 48, 82, 89–90
  1926 BAAS meeting in 25, 49, 83
  University of 8, 183, 184, 186, 238
Owen, Richard 140, 158

Palestine Exploration Fund 71, 72, 79, 99n.13
Phillips, John 31, 33–4, 142
physics 70
  see also section A (mathematics and physics)

Index 277

Plymouth
  1841 BAAS meeting in 25, 47
  1877 BAAS meeting in 25, 48, 81
polar science 8, 66, 71, 82, 84, 143–4, 159
Portsmouth
  1911 BAAS meeting in 25, 49, 83, 204–9, 241
practice, science as 36–44, 59–61, 95–7
  see also *conversazione*; display; excursions; lecturing
Prichard, James Cowles 168–9, 173–4, 178, 181
Punch 51, 52

Queries respecting the Human Race, to be addressed to Travellers and Others 169–71, 177, 181–2, 217–18

race 166, 168–9, 178, 192, 215–16
Ravenstein, E. G. 71, 72, 73, 82, 117, 119, 185
regional survey 216–21, 230n.62
Rivett, A. R. 110, 118, 124, 132n.34
Ross, Sir James Clark 142–4, 146, 150, 158, 169
Ross, Sir John 56, 145
Rougemont, Louis de 190, *191*
Roxby, P. M. 75, 84, 187, 195n.34, 215
Royal Geographical Society 7–8, 66, 68, 79, 91, 92, 97, 104, 138–9, 165–6, 172–4, 183–5, 205, 221–3, 240–2
Royal Scottish Geographical Society 8, 91, 101n.37, 186, 210–11
Royal Society 69, 143, 147, 148, 165, 207, 212, 213
Rudmose-Brown, R. N. 76, 84, 187, 210
Russell, Sir John 75, 76, 218

Sabine, Edward 142–3, 146, 148–9, 158
science
  as practice 36–44, 59–61, 95–7
  see also *conversazione*; display; excursions; lecturing
  civic, idea of xv, 3–9, 24–61, 233–7
  colonialism and 9–10, 103–29, 213–16
  empire and 9–11, 79–80, 82–3, 103–29, 199–200
  geography of xv, 1–16, 24–61, 68–77, 232–45
  history of xv, 1–16, 232–45
  spatial turn in historical studies of 2, 10–16, 232–45
'scientific geography', idea of 77–89
Scottish Geographical Magazine 91, 221, 222
section A (mathematics and physics) 15, 56, 69, 119, 121, 147, 202, 204, 236
section B (chemistry) 69, 202, 203
section C (geology) 12, 69, 76, 138–9, 189, 200, 202, 236
section D (biology) 138, 236
section D (botany) 55, 68, 69, 76, 137, 202, 236, 239
section D (zoology) 76, 202, 236
section E (geography)
  Committee on the Human Geography of Tropical Africa 75, 119, 181–2
  Committee on the Teaching of Geography 71–2, 213–14
  funding of 68–77, 79–80, 95–6, 135, 137, 187, 240
  grants awarded to 69, 72–6, 152
  history of as a section xv, 2–3, 10–11, 66–97, 168–77, 200–12, 223, 238–42
  in BAAS overseas meetings xv, 103–29
  presidential addresses 12, 29, 36, 80–4, 100n.22, 111, 120–1, 187–8, 236
  recommendations proposed by 76–7, 152–3
  relationship of with section C (geology) 12, 69, 200–4
  relationship of with section H (anthropology) 69, 119, 168–77, 202, 238–9
  relationship of with section L (education) 69, 202, 213, 223–4

section E (geography) (*cont.*)
  seen as the 'Ladies' section' 15, 93, 175, 177
  working practices of 3, 50, 66–7, 69, 77–89, 103–29
  see also geography
section F (economics) 15, 69, 117, 202
section G (engineering) 69, 202, 235
section H (anthropology) 68, 69, 76, 118, 119, 124, 172–7, 185, 188, 189, 202, 240
section I (physiology) 69, 124, 202
section L (education) 15, 69, 119, 121, 202, 213, 223
section K (botany) 76, 117, 202, 236, 245n.7
section M (agriculture) 69, 76, 202
Sheffield
  1879 BAAS meeting in 25, 36, 48, 56, 81, 86, 155
  1910 BAAS meeting in 25, 49, 83, 204
Simpson, J. Y. 103, 131n.17
South Africa
  1905 BAAS meeting in 24, 25, 41, 48, 80, 103, 105, 107–8, 121, 126
  1929 BAAS meeting in 24, 25, 41, 49, 80, 84, 103, 105, 108, 118–19, 133n.49, 181
South African Association for the Advancement of Science 107, 108, 109
Southampton
  1846 BAAS meeting in 25, 47
  1882 BAAS meeting in 25, 48, 82
  1925 BAAS meeting in 25, 49, 83, 218
Southport
  1883 BAAS meeting in 25, 47, 82
  1903 BAAS meeting in 25, 47, 82, 217
speech
  as a performance in science 12–13, 14–15, 36, 38, 92–4, 236–7
  see also lecturing; section E (geography), presidential addresses
Stanley, Henry Morton 51, 53, 80, 92, 94, 153–4, 185, 243
Strachey, Lt. Gen. Richard 81, 85, 90, 184

surveying, 8, 14, 71, 79, 80, 149–50, 154
  see also mapping; regional survey
Swansea
  1848 BAAS meeting in 25, 30, 47
  1880 BAAS meeting in 25, 48, 81
Sydney 108, 110, 118, 235

Taylor, Griffith 84, 114
terrestrial magnetism 82, 136, 140–50, 165, 227n.20
Thackray, Arnold 9–10, 24–5, 26, 55, 136
Thomson, Charles Wyville 80, 81
Times, The 42, 56, 108, 129, 206
Toronto
  1897 BAAS meeting in xv, 24, 25, 48, 79, 82, 103, 106, 116–17, 119, 120–1, 123
  1924 BAAS meeting in 24, 25, 49, 79, 83, 103, 114, *115*, 118, *123*, *125*, 126
Tylor, E. B. 181, 185
Tyndall, John 12, 152

Wallace, Alfred Russel 92, 153, 176
Whewell, William 16n.3, 32, 142, 239
Wilson, Sir Charles W. 71, 72, 81
Winnipeg
  1909 BAAS meeting in 24, 25, 49, 79, 83, 103, 112, 117–18, 234
women
  in BAAS audiences 47–9, 55–7, 95, 124, 171, 175, 234
  see also audiences; section E (geography), seen as the 'Ladies' section'

York
  1831 BAAS meeting in 24, 25, 47, 139
  1844 BAAS meeting in 25, 47
  1881 BAAS meeting in 25, 48, 80–1
  1906 meeting in 25, 48, 83–4, 87

Zambesi Expedition (1858–63) 85, 92
zoology 41, 55, 76, 118, 149, 155, 156, 217, 239
  see also section D (zoology)

EU authorised representative for GPSR:
Easy Access System Europe, Mustamäe tee 50,
10621 Tallinn, Estonia
gpsr.requests@easproject.com

www.ingramcontent.com/pod-product-compliance
Lightning Source LLC
Chambersburg PA
CBHW021820300426
44114CB00009BA/256